Springer Theses

Recognizing Outstanding Ph.D. Research

Aims and Scope

The series "Springer Theses" brings together a selection of the very best Ph.D. theses from around the world and across the physical sciences. Nominated and endorsed by two recognized specialists, each published volume has been selected for its scientific excellence and the high impact of its contents for the pertinent field of research. For greater accessibility to non-specialists, the published versions include an extended introduction, as well as a foreword by the student's supervisor explaining the special relevance of the work for the field. As a whole, the series will provide a valuable resource both for newcomers to the research fields described, and for other scientists seeking detailed background information on special questions. Finally, it provides an accredited documentation of the valuable contributions made by today's younger generation of scientists.

Theses are accepted into the series by invited nomination only and must fulfill all of the following criteria

- They must be written in good English.
- The topic should fall within the confines of Chemistry, Physics, Earth Sciences, Engineering and related interdisciplinary fields such as Materials, Nanoscience, Chemical Engineering, Complex Systems and Biophysics.
- The work reported in the thesis must represent a significant scientific advance.
- If the thesis includes previously published material, permission to reproduce this must be gained from the respective copyright holder.
- They must have been examined and passed during the 12 months prior to nomination.
- Each thesis should include a foreword by the supervisor outlining the significance of its content.
- The theses should have a clearly defined structure including an introduction accessible to scientists not expert in that particular field.

More information about this series at http://www.springer.com/series/8790

Yuki Harada

Interactions of Earth's Magnetotail Plasma with the Surface, Plasma, and Magnetic Anomalies of the Moon

Doctoral Thesis Accepted by
Kyoto University, Kyoto, Japan

Springer

Author (Present address)
Dr. Yuki Harada
Space Sciences Laboratory
University of California
Berkeley, CA
USA

Supervisor
Dr. Akinori Saito
Department of Geophysics
Kyoto University
Kyoto
Japan

ISSN 2190-5053 ISSN 2190-5061 (electronic)
ISBN 978-4-431-56256-6 ISBN 978-4-431-55084-6 (eBook)
DOI 10.1007/978-4-431-55084-6

Springer Tokyo Heidelberg New York Dordrecht London

Printed on acid-free paper

Springer is part of Springer Science+Business Media (www.springer.com)

Parts of this thesis have been published in the following journal articles:

- Harada, Y., S. Machida, Y. Saito, S. Yokota, K. Asamura, M. N. Nishino, T. Tanaka, H. Tsunakawa, H. Shibuya, F. Takahashi, M. Matsushima, and H. Shimizu (2010), Interaction between terrestrial plasma sheet electrons and the lunar surface: SELENE (Kaguya) observations, Geophys. Res. Lett., 37, L19202, doi: 10.1029/2010GL044574.
- Harada, Y., S. Machida, Y. Saito, S. Yokota, K. Asamura, M. N. Nishino, H. Tsunakawa, H. Shibuya, F. Takahashi, M. Matsushima, and H. Shimizu (2012), Nongyrotropic electron velocity distribution functions near the lunar surface, J. Geophys. Res., 117, A07220, doi: 10.1029/2012JA017642.
- Harada, Y., S. Machida, J. S. Halekas, A. R. Poppe, and J. P. McFadden (2013), ARTEMIS observations of lunar dayside plasma in the terrestrial magnetotail lobe, J. Geophys. Res., 118, 3042–3054, doi: 10.1002/jgra.50296.
- Harada, Y., S. Machida, Y. Saito, S. Yokota, K. Asamura, M. N. Nishino, H. Tsunakawa, H. Shibuya, F. Takahashi, M. Matsushima, and H. Shimizu (2013), Small-scale magnetic fields on the lunar surfaceinferred from plasma sheet electrons, Geophys. Res. Lett., 40, 3362–3366, doi: 10.1002/grl.50662.
- Harada,Y., Y. Futaana, S. Barabash, M. Wieser, P. Wurz, A. Bhardwaj, K. Asamura, Y. Saito, S. Yokota, H. Tsunakawa, S. Machida (2014), Backscattered energetic neutral atoms from the Moon in the Earth's plasma sheet observed by Chandarayaan-1/Sub-keV Atom Reflecting Analyzer instrument, J. Geophys. Res., 119, 3573–3584, doi: 10.1002/2013JA019682.

Supervisor's Foreword

The importance of understanding the Moon–plasma interaction has been recognized since the early days of space age not only because it contains fundamental and fascinating plasma physics but also because measurements of lunar electromagnetic environment raised concerns for its possible impacts on human and robotic exploration on the lunar surface. In this thesis, Yuki Harada has analyzed a wide variety of data from recent lunar missions (i.e., Kaguya, ARTEMIS, and Chandrayaan-1) to investigate the basic aspects of lunar plasma environment in the Earth's magnetotail, where the Moon is bathed in a hot plasma of terrestrial plasma sheet or in an extremely tenuous plasma of magnetotail lobes. The Moon–plasma interactions in these plasma regimes make a clear contrast with those in the solar wind, which consist of relatively cold, supersonic plasma.

From Kaguya observations, Dr. Harada reveals that nongyrotropic electron velocity distribution functions are formed by electron absorption by the lunar surface combined with the electron gyromotion. This is a unique observation because usually it is difficult to detect nongyrotropic signatures of electrons in space plasmas. In the hot plasma of the Earth's plasma sheet, a typical electron gyrodiameter becomes so large that Kaguya can measure the nongyrotropic electrons even at its nominal altitude of 100 km. Dr. Harada conducts test particle simulations to compare with these electron observations, taking into account electric and magnetic fields around the Moon. He also utilizes nonadiabatic magnetic scattering of high-energy electrons by lunar crustal magnetization to infer small-scale magnetic fields on the lunar surface.

Dual-probe ARTEMIS observations, on the other hand, exhibit dominance of plasma of lunar origin over the ambient plasma when the Moon is in the Earth's magnetotail lobes. The lobe plasma is so tenuous that the lunar plasma density can be several times higher than the lobe plasma density. This present work suggests the significant role in modifying the lobe plasma played by the ionized atmosphere of the Moon, which is often thought of as an "airless" body.

Finally, Dr. Harada analyzed energetic neutral atom (ENA) data from Chandrayaan-1, showing that plasma sheet protons are backscattered at the lunar surface as hydrogen ENAs. The ENA observations together with test particle simulations suggest that the lunar mini-magnetosphere, which results from the solar–wind interaction with the strongly magnetized region on the lunar surface, may "disappear" in the Earth's plasma sheet owing to the less effective magnetic shielding from the hotter protons.

I believe that the results presented in this thesis are an outstanding contribution to the literature and are relevant for understanding interactions of generally hot plasma with the Moon as well as other airless bodies in the solar system and beyond.

Kyoto, May 2014 Dr. Akinori Saito

Acknowledgments

I would like to express my heartfelt gratitude for the guidance and support of my former thesis advisor, Prof. Shinobu Machida. I am grateful for all his contributions throughout this thesis both at Kyoto University and after he moved to Nagoya University. I also gratefully acknowledge my current advisor, Dr. Akinori Saito, for his generous support to complete this thesis. In addition, I express my appreciation to Emeritus Prof. T. Araki, Prof. T. Iyemori, Dr. H. Toh, Dr. M. Takeda, Dr. M. Nosé, and other colleagues at the Solar-Planetary Electromagnetism Laboratory and World Data Center for Geomagnetism, Kyoto. Their encouraging and constructive comments have made my Ph.D. experience productive and stimulating.

I deeply thank Dr. Y. Saito, Dr. S. Yokota, Dr. K. Asamura, Dr. M.N. Nishino, Prof. H. Tsunakawa, Prof. H. Shibuya, Dr. F. Takahashi, Dr. M. Matsushima, Dr. H. Shimizu, and other MAP-PACE and MAP-LMAG team members for their great support in processing and analyzing the MAP data. I am also grateful to the system members of the SELENE project.

To Dr. J.S. Halekas, Dr. A.R. Poppe, Dr. J.P. McFadden, and other colleagues at the Space Sciences Laboratory at U.C. Berkeley, I am thankful for the opportunity to visit and collaborate with their NASA lunar science institute team on ARTEMIS data analysis. I acknowledge NASA contract NAS5-02099 and V. Angelopoulos for use of data from the THEMIS Mission; J.W. Bonnell and F.S. Mozer for use of EFI data; A. Roux and O. LeContel for use of SCM data; and K.H. Glassmeier, U. Auster, and W. Baumjohann for use of FGM data provided under the lead of the Technical University of Braunschweig and with financial support through the German Ministry for the Economy and Technology and the German Center for Aviation and Space (DLR) under contract 50 OC 0302.

I would like to express my sincere thanks to Dr. Y. Futaana, Prof. S. Barabash, and other colleagues at the Swedish Institute of Space Physics for their great support for Chandrayaan-1 data analysis and my daily life in Kiruna during my visit.

This work was supported by a Research Fellowship for Young Scientists, awarded by the Japan Society for the Promotion of Science.

Contents

Chapter 1
Introduction

Abstract In this chapter, I provide extended background information on the Moon–plasma interaction. It begins with a brief outline of plasma interactions with solar system bodies. Key parameters controlling the nature of interactions are the body's atmospheric density and the strength of its intrinsic magnetic field. I then introduce particle emission processes from solid surfaces of airless bodies in space environment. Particle emission and surface charging are fundamental consequences of direct interactions of airless bodies with the surrounding plasma. Next I present a review of observational and theoretical studies of lunar plasma environment. Our current understanding of the Moon–plasma interaction is mostly based on a number of plasma and electromagnetic field measurements which have been conducted since the Apollo era. I also describe the basic characteristics of the lunar tenuous exosphere. Finally I mention the objectives of this thesis.

Keywords Plasma interactions with solar system bodies · Solid surfaces in space · Lunar plasma environment

1.1 Plasma Interactions with Solar System Bodies

Flowing plasma interactions with various types of bodies can be drastically different from one body to another, depending on the strength of a possible intrinsic magnetic field and on the density of the atmosphere and ionosphere [9, 148]. In terms of characteristics of the plasma interactions, the solar system bodies can be categorized into four groups (cf. Table 1.1): (i) bodies with both strong magnetic fields and dense atmospheres (e.g., Earth, Jupiter, Saturn, Ganymede), (ii) bodies that have dense atmospheres but lack strong magnetic fields (e.g., Venus, Mars, Titan, comets close to the Sun), (iii) bodies that have strong magnetic fields but lack dense atmospheres (e.g., Mercury), and (iv) bodies that lack both strong magnetic fields and dense atmospheres (e.g., the Earth's moon, asteroids, distant comets). In this section, I briefly outline the interactions between bodies in the solar system and their surrounding plasma.

Y. Harada, *Interactions of Earth's Magnetotail Plasma with the Surface, Plasma, and Magnetic Anomalies of the Moon*, Springer Theses,
DOI 10.1007/978-4-431-55084-6_1

Table 1.1 Classes of plasma interactions with solar-system bodies

	Dense atmosphere	Tenuous atmosphere
Strong magnetic field	Earth, Jupiter, Saturn, Uranus, Neptune	Mercury, Ganymede
Weak magnetic field	Venus, Mars, Titan, near-Sun comets	Moon, asteroids, Phobos, Tethys, Rhea, distant comets

1.1.1 Bodies with Intrinsic Magnetic Fields and Dense Atmospheres

A planetary magnetic field, produced by a presently active internal dynamo, acts as an effective obstacle to the solar wind plasma [16, 17]. The solar wind dynamic pressure compresses the dayside planetary magnetic field and confines the nightside filed into a long tail stretching anti-sunward. In order to deflect the supersonic solar wind plasma, the fast-mode magnetosonic wave becomes nonlinear and forms a bow shock. The solar wind is slowed and heated at the shock and becomes subsonic in the subsolar magnetosheath. The magnetosphere of a dense-atmosphere planet, extended far beyond the dense ionosphere, prevents the solar wind plasma from directly interacting with the ionosphere except the small "cusp" regions at the north and south poles. Precipitation of the solar and magnetospheric electrons and ions into the upper atmosphere causes auroral emission. The shapes and sizes of the magnetospheres in the solar wind depend on a number of parameters such as the planetary magnetic moments, solar-wind dynamic pressure, interplanetary magnetic field (IMF), and magnetospheric plasma pressure.

1.1.2 Unmagnetized Bodies with Dense Atmospheres

Unmagnetized bodies with substantial atmospheres, such as Mars, Venus and comets near the Sun, possess ionospheres formed mainly by photoionization of the atmospheric neutral particles by the solar photons. Since the ionosphere is a conductor, the convecting magnetic field induces electric currents in it, preventing a bulk penetration of the magnetic field by generating a canceling field. The supersonic solar-wind plasma, frozen to the IMF, must be diverted around the conducting obstacle, forming a bow shock. The IMF piles up on the dayside and drapes around the body. This magnetic barrier screens the ionosphere from the solar wind plasma. Such an "induced magnetosphere" is a common feature of the interaction of the solar wind with ionospheres of unmagnetized bodies [21, 27, 114]. On the other hand, Titan, the Saturn's largest moon with a dense nitrogen-rich atmosphere, is immersed most of the time in the Saturn's magnetospheric plasma flowing at subsonic speed. In this case, the magnetic field still piles up on the upstream side, but no bow shock is formed [6, 93].

1.1.3 Airless Bodies with Intrinsic Magnetic Fields

Mercury does not have a significant atmosphere, but does possess a moderate intrinsic magnetic field, creating a magnetosphere similar in shape but smaller than the Earth's [2, 3, 89, 119]. The absence of a thick atmosphere leads to direct interactions of the planetary solid surface with the magnetospheric particles, and with the solar wind particles in the cusp regions. The cusp regions of Mercury are rather large because of the small magnitude of its dipole field, and the solar-wind sputtering in the cusp regions is considered as a possible source of planetary ions and neutrals [100, 188, 196]. The magnetosphere of Jupiter's moon, Ganymede, is located inside the jovian magnetosphere, interacting with the jovian plasma and magnetic field corotating at subsonic speed [97, 98].

1.1.4 Airless Bodies Without Intrinsic Magnetic Fields

A body like our moon, which is composed of essentially insulating material and lacks both a significant atmosphere and an intrinsic magnetic field, is directly exposed to its surrounding plasma. The majority of the impinging plasma is absorbed or neutralized by the surface material and the magnetic field rapidly diffuses into the body's interior. In this sense, the body can be thought of as a plasma absorber, causing no significant pressure perturbation in the upstream plasma. Therefore, even if the body is immersed in the supersonic solar wind, no bow shock is formed upstream. In the downstream region, a plasma void is left behind the body and the surrounding plasma expand into the body's "wake". The pressure gradient at the wake boundary produces the diamagnetic current system, causing magnetic perturbations in the downstream region in both supersonic and subsonic plasma flows [38, 58, 95, 121, 147, 162]. The plasma environment in the vicinity of the Moon is introduced in more detail in the following sections.

1.2 Solid Surfaces in Space Environment

When energetic photons, charged and neutral particles are incident on a solid surface, a portion of their energies can be transferred to other particles, which may escape from the surface. The incident particles may be absorbed by the surface material or scattered back into space. As a result, charged and neutral particles are emitted from the bombarded surface with different characteristics depending on the emission mechanisms. Also, absorption and emission of charged particles give rise to surface charging. The near-lunar environment provides a unique natural laboratory to study the surface-plasma interactions, because the surface of the Moon, which lacks both a significant atmosphere and an intrinsic magnetic field, is directly exposed to the ambient environment and the emitted particles can be detected by spacecraft far above the surface. This section contains a brief review of particle emission mechanisms from a solid surface, particularly focusing on the energy characteristics of the emitted

Table 1.2 Particle emission processes from a solid surface in space

Process	Incident particle	Emitted particle	Typical energy of emitted particle
Photoemission	Photon (under solar irradiation)	Electron	A cold core (a few eV) and a high-energy tail (>~10 eV)
Backscattering	Electron	Electron	A fraction of the incident energy
	Ion	Neutral atom, ion	A fraction of the incident energy
Secondary electron emission	Electron, ion	Electron	A few eV
Sputtering	Photon, electron, ion	Neutral atom, ion	~0.1–10 eV

particles (the emission processes are summarized in Table 1.2), as well as a short introduction of charging processes of solid surfaces.

1.2.1 Photoemission

Incident photons with energies above the work function of the surface material can cause photoelectron emission. Photoemission properties were measured in laboratory for typical materials such as aluminum, gold, stainless steel and graphite, with the surfaces treated so as to simulate space conditions [41]. The measured work functions range from 4.0 to 4.8 eV. On the other hand, the work function of the lunar regolith was experimentally studied in a similar way and determined to be 5 eV [40]. Typically, the photoemission yield, defined as the number of emitted photoelectrons per incident photon, peaks at ~10–20 eV, which is in the ultraviolet (UV) and extreme ultraviolet (EUV) ranges.

Energy spectra of the emitted photoelectrons are determined by the product of the incident photon spectrum and the photoemission yield as a function of energy. The solar flux spectrum falls off sharply by five orders of magnitudes for photon energies between 4 and 10 eV, and is flatter for higher energies [56]. In general, photoelectron energy spectra under solar irradiation in space consist of two populations: a cold core and a high-energy tail. The cold core distribution can be approximated by the Maxwellian distribution with characteristic temperatures of a few eV [41, 56]. The high-energy tail population is not yet well-understood, but its existence is inferred from large positive potentials of spacecraft surfaces measured in the terrestrial magnetotail lobes [131].

1.2.2 Backscattering

When energetic electrons and ions impinge on a surface, they can penetrate the surface and collide with lattice electrons and atoms in the material [110, 183]. After the collisions, some may eventually backscatter out of the material into space, losing a fraction of their initial energies. The electron backscattering coefficient depends on the incident electron velocity and on the surface material, and resulting net albedo for ~100 eV isotropic electrons is around ~10–30 % [94]. The majority of the impinging ions are expected be neutralized by resonant or Auger neutralization before absorbed or backscattered [123, 187]. A small portion, less than a few %, of the incident ions are re-emitted as ions [154, 183]. These backscattered particles have similar but somewhat lower energies than their initial energies.

1.2.3 Secondary Electron Emission

A part of the energy and momentum that an impinging particle loses when colliding in the surface material can go into exciting surface electrons, causing secondary electron emission. The secondary emission yield depends on the incident particle energy and the surface material, and it can be greater than unity in some cases [94, 183]. The higher-energy incident particles can excite more electrons in the material, but they penetrate deeper in the material. The electrons excited deeper below the surface are less likely to escape from the surface. Consequently, the secondary emission yield functions have peaks at primary energies of several hundreds eV for electron impact, and of ~10–100 keV for proton impact. The energy distributions of the secondary electrons can be represented by the Maxwellian with characteristic energies of a few eV.

1.2.4 Sputtering

Sputtering was originally defined as the removal of atoms form a surface by ion impact, but the definition is often broadened to include the removal of atoms by the impact of incident photons, neutral atoms, and electrons [110]. Photon sputtering, also known as photon-stimulated desorption, is emission of neutral particles from the surface caused by the injection of the incident photon energies [111]. Laboratory studies show that the energy distribution of the neutral atoms emitted by photon sputtering corresponds to characteristic temperatures of ~800–2000 K (0.07–0.17 eV) [164, 171, 185]. On the other hand, charged particle sputtering produces a Thompson-Sigmund distribution with peaks at ~1–10 eV [110, 161, 170, 189]. A few percent of the sputtered atoms are emitted as ions (also called secondary ions) [30, 190].

1.2.5 Surface Charging

Surface charging of an airless body in space has been extensively studied, specifically with the aim of preventing spacecraft malfunctions attributed to spacecraft charging and electrostatic discharges [183]. Surface absorption of incident charged particles, as well as charged particle emission caused by impinging high-energy particles, results in charge transfer between the surface and space. The electric field generated by the charge that has already accumulated on the surface attracts or repels the incident charged particles, changing the charge-transfer rate. This process continues until it reaches an equilibrium that the net current onto the surface balances. In general, the equilibrium potential will not be zero. If the photoelectron current is absent (i.e., surfaces in optical shadow), electron thermal currents generally dominate owing to higher mobility of electrons than heavier ions, resulting in negative surface potentials on the order of the ambient electron temperature. Energetic electrons in the Earth's magnetosphere cause negative potentials on geosynchronous satellites as large as several kV during eclipse [25].

The surface equilibrium potential will be positive in regions where the photoelectron current dominates. Considering typical photoelectron temperatures of a few eV, a straightforward charging theory predicts slightly positive potentials of sunlit surfaces of less than +10 V. However, in the terrestrial magnetotail lobes, where the incident electron current from the ambient plasma becomes extremely small, the surface potential can become much higher than expected from a simple current-balance model based on Maxwellian plasmas. In this region, the energy distribution of the photoelectrons has important implications for the surface potential. Pederson [131] suggested that a high-energy tail population in the photoelectron energy distribution causes the large positive spacecraft potentials of tens of volts detected in the Earth's magnetotail. The spacecraft surface potentials can reach approximately +70 V in the tail lobes [158].

1.3 Lunar Plasma Environment

The Moon orbits the Earth with an orbital radius of $\sim 60 R_E$ ($1 R_E = 6{,}378$ km). Along its way, it passes through various regions, including the solar wind, the terrestrial magnetosheath, the magnetotail lobes, and the plasma sheet as sketched in Fig. 1.1. In each of these regions, the plasma conditions near the Moon vary widely in density and energy. The various types of plasma interact directly with the lunar surface because of the lack of both a global magnetic field and a thick atmosphere. In this section, I present an overview of past observations and models regarding the lunar plasma environment in the solar wind and the Earth's magnetotail to provide context for this thesis.

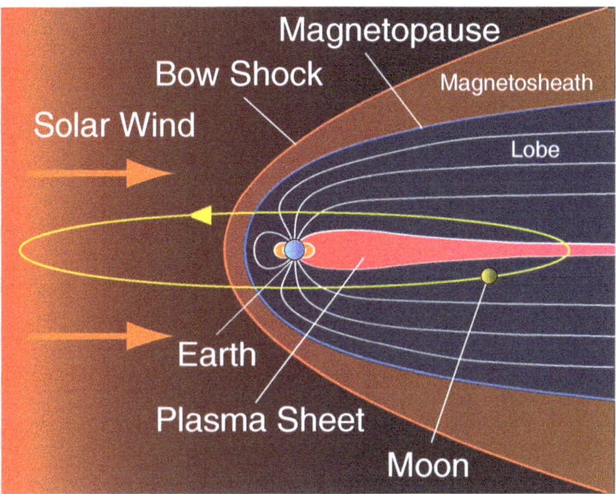

Fig. 1.1 Schematic illustration of the lunar plasma environment

1.3.1 Lunar Wake in the Solar Wind

As summarized in Table 1.3, the lunar electromagnetic environment has been studied since plasma and magnetic fields were first measured in the 1960s and 1970s [159]. Explorer 35 observations revealed the fundamental features of the interaction between the Moon and the solar wind. No bow shock is formed in the upstream solar wind [120], while a tenuous region, called the lunar wake, is formed behind the Moon because of the absorption of the solar-wind plasma by the lunar surface [104]. Magnetometer measurements exhibited enhanced magnetic fields in the wake as well as reduced fields near the wake boundary, which can be explained in terms of diamagnetic-current systems [22, 23, 121]. A diamagnetic sheet current at the wake boundary, caused by the decrease in density of gyrating particles, produces a magnetic field, which increases in the central wake. The magnetic signatures in the lunar wake were successfully described by several theoretical models [90, 112, 181, 182].

Measurements conducted by several spacecraft in the 1990s revealed another characteristic signature of the lunar wake, namely ambipolar electric fields near the wake boundary. An electrostatic potential drop across the wake boundary is formed by charge separation between faster electrons expanding into the void ahead of the slower ions [155]. Data from a lunar wake passage by the Wind spacecraft displayed the existence of ion beams parallel to the magnetic field refilling the wake from sides, inferring the presence of an ambipolar potential drop [130]. When the magnetic field line passing through spacecraft was connected to the lunar wake, NOZOMI detected counter-streaming electrons, which were presumably reflected by the wake potential [51], and WIND and GEOTAIL observed both electrostatic and electromagnetic waves associated with the modified velocity distribution functions (VDFs) of the solar wind particles [7, 31, 118]. In addition to the data from these lunar flybys,

Table 1.3 Plasma and electromagnetic field measurements on and above the Moon

Mission	Year	Instrument	Altitude	Inclination
Lunar 2	1959	Magnetometer	>50 km	(impact)
Lunar 10	1966	Magnetometer, charged particle detector	350–1015 km	72°
Explorer 35	1967–1973	Magnetometer, Faraday cup plasma detector	1.4–5.4 R_M	11°
Apollo 12 landing site	1969–1977	Magnetometer, Faraday cup plasma detector, electrostatic ion analyzer	Surface	–
Apollo 14 landing site	1971–1977	Magnetometer, electrostatic ion analyzer, electrostatic electron and ion analyzer	Surface	–
Apollo 15 landing site	1971–1977	Magnetometer, Faraday cup plasma detector, electrostatic ion analyzer	Surface	–
Apollo 15 subsatellite	1971	Magnetometer, electrostatic electron analyzer, solid state telescope (electron, ion)	Initially 100, 40–200 km (gravity perturbation)	28°
Apollo 16 landing site	1972–1977	Magnetometer	Surface	–
Apollo 16 subsatellite	1972	Magnetometer, electrostatic electron analyzer, solid state telescope (electron, ion)	Initially 100, >0 km (crash to the Moon)	10°
Lunokhod-2	1973	Magnetometer	Surface	–
WIND	1994	Magnetometer, electric field sensor, Faraday cup plasma detector, electrostatic electron and ion analyzer	6.5 R_M	(flyby)
GEOTAIL	1994	Magnetometer, electric field sensor, electrostatic electron and ion analyzer	27 R_M	(flyby)
NOZOMI	1998	Electrostatic electron and ion analyzer	1.6 R_M	(flyby)
Lunar prospector	1998–1999	Magnetometer, electrostatic electron analyzer	Nominal 100 km, extended >0 km	90°
Kaguya	2007–2009	Magnetometer, electric field sensor, electrostatic electron and ion analyzer	Nominal 100 km, extended >0 km	90°
Chang'E-1	2007–2009	Electrostatic ion analyzer	100–200 km	90°
Chandrayaan-1	2008–2009	Electrostatic ion analyzer, energetic neutral atom analyzer	100–200 km	90°
Chang'E-2	2010–2011	Electrostatic ion analyzer	Nominal 100 km	90°
ARTEMIS	2010–present	Magnetometer, electric field sensor, electrostatic electron and ion analyzer, solid state telescope (electron, ion)	~100–19,000 km	~10–20°

$1 R_M = 1{,}738$ km

three-dimensional (3D) measurements of low-energy electrons and magnetic fields were conducted by the Lunar Prospector at a low-altitude polar orbit ($<\sim 100$ km above the Moon).

A large amount of data obtained by Lunar Prospector enabled a statistical study on the global structure of the lunar wake, which indeed demonstrated the existence of magnetic perturbations by the diamagnetic current system as well as of negative electrostatic potentials in the wake [58]. More sophisticated numerical models than the early work in the Apollo era have been developed to investigate the fluid and kinetic features of the lunar wake [12, 14, 15, 32, 34, 91, 96].

Since the late 2000s, several lunar orbiters have provided whole new datasets of plasma and fields in the vicinity of the Moon [73]. Kaguya conducted the first ion measurements by modern instruments at low-altitude orbits ($<\sim 100$ km) [152, 153]. The ion data obtained by Kaguya in the near-lunar wake revealed two kinetic processes of perpendicular ion entry into the wake: gyration of solar wind ions combined with the ambipolar electric fields near the wake boundary, resulting in pair-wise energy gain/loss features observed at the two flanks of the wake (Type-I entry) [128], and self-pickup of solar wind ions backscattered from the dayside surface, allowing ion entry deep into the near-lunar wake (Type-II entry) [125]. The excess ion populations caused by the latter mechanism accelerate counter-streaming electron beams into the wake, generating broadband electrostatic noise [126]. In addition, Chandrayaan-1, orbiting at $\sim$100–200 km altitudes, detected ion entry into the deep wake along the field line [48], and perpendicular entry to the field line in the case of IMF parallel to the solar wind flow [26]. These ion observations at low altitudes in the deep wake, which cannot be explained by the conventional fluid models and demonstrate essentially kinetic nature of plasma expansion processes into the lunar wake, have driven a series of theoretical and numerical investigations [37, 38, 81, 82, 85, 86, 88, 115, 175]. The kinetic ion entry into the lunar wake may explain the Apollo Suprathermal Ion Detector Experiment (SIDE) measurements of the solar wind ions on the lunar surface in the deep night [43] Recent distant-wake observations by ARTEMIS have promoted further understanding of the global and local features of the lunar wake [169, 184, 192].

1.3.2 Solar–Wind Interactions with the Dayside Lunar Surface

It has been thought that impinging charged particles are almost completely absorbed in the porous regolith surface of the Moon, but recent observations from Kaguya, Chandrayaan-1 and IBEX indicate that $\sim$0.1–1 % of the incident solar wind protons are scattered back as ions [154] and $\sim$10 % as energetic neutral atoms (ENAs) [106, 187]. The backscattered protons are accelerated via "self-pickup" process, resulting in the maximum energy up to nine times solar wind energy [153, 154]. The backscattered protons interact with the upstream solar wind plasma, generating whistler and magnetohydrodynamic waves [72, 116, 117]. As for the backscattered hydrogen atoms, the magnetic and electric fields no longer affect the trajectories of these ENAs, which are approximated by straight lines. The observed energy spectra of the backscattered hydrogen ENAs are relatively flat up to hundreds of electronvolts and decrease sharply at higher energies [1, 46, 49, 146]. Angular distributions of the backscattered ENAs are almost isotropic for normal incidence of the solar wind, and exhibit anisotropy for large solar zenith angles [157].

1.3.3 Solar–Wind Interactions with Lunar Magnetic Anomalies

Interest in solar–wind interactions with small-scale magnetic fields has been spurred by the discovery of lunar crustal magnetism and its correlations with the occurrence of magnetic "limb compressions" in the Apollo era [28, 45, 121, 149, 159, 163]. The analysis of electron and magnetic field data from Lunar Prospector suggested the formation of a "mini-magnetosphere" as a result of solar-wind interaction with lunar magnetic anomalies [101]. Upward ions with extremely large flux up to more than 10 % of the solar wind proton flux are observed above magnetic anomalies [103, 153]. Several spacecraft and surface observations as well as numerical simulations indicate that solar wind protons are reflected or deflected above the surface not only by crustal magnetic fields, but also by charge-separation electric fields and magnetic fields associated with current systems around magnetic anomalies [8, 20, 50, 75, 92, 103, 122, 134, 151]. The solar-wind interaction with magnetic anomalies is characterized by the upstream solar wind parameters such as the solar wind dynamic pressure, Mach numbers, strength and polarity of the interplanetary magnetic field [29, 61, 74, 99, 127, 177]. Various kinds of electrostatic and electromagnetic waves, presumably associated with the magnetically reflected ions and electrons, are observed above the magnetic anomalies [59, 60, 71, 76, 101, 172, 173].

An important consequence of the formation of the mini-magnetosphere is that the lunar surface might be shielded from the impinging ions. Several authors pointed out correlations between the magnetic anomalies and high-albedo markings on the lunar surface (the lunar "swirls"), and it is proposed that swirl origins are related to differential space weathering caused by magnetic shielding from the solar wind protons and/or charged-dust transport by electric fields associated with the mini-magnetosphere [13, 52, 77, 83, 84, 144, 145, 179]. The surface distribution of precipitating ion flux is the essential information for the understanding of the ion shielding processes and their relations with the lunar swirls. However, only a few measurements on the lunar surface of the solar wind ions exist [159].

Lunar ENA observations can be used as a remote sensing technique to estimate the precipitating ion flux onto the surface, and consequently, to image the mini-magnetosphere [47]. If a local region on the lunar surface is completely shielded from the incident ions by the magnetic field, no ENAs caused by ion precipitation will be emitted from that region. In fact, a reduction in the ENA flux was observed by CENA above magnetic anomalies, indicating locally effective ion shielding [177, 186]. The CENA observations also revealed an enhanced-flux region surrounding the void region, suggesting increased solar wind flux due to the deflection around the mini-magnetosphere. Recently, global ENA maps were produced from the CENA data, which show large reductions in ENA backscattering ratio (defined as the ratio of the upward ENA flux to the downward flux of the upstream solar-wind protons) as much as $\sim$50 % in large and strong anomaly regions [178]. In addition, the lunar ENA observations were used to infer a potential difference between the upstream plasma and the lunar surface, which is potentially associated with charge separation above the magnetic anomalies [50].

1.3.4 Moon–Plasma Interactions in the Earth's Magnetotail

While the Moon–solar wind interaction has been extensively investigated, relatively little work has addressed Moon-plasma interactions in the Earth's magnetotail. During the Apollo era, lunar shadowing techniques were used to infer magnetospheric convection electric fields at the Moon's distance [4, 176]. The presence of convection electric fields can be deduced through detection of the $\mathbf{E} \times \mathbf{B}$ drift displacement of low-energy electrons at the edge of the lunar electron shadow. A typical value of 0.15 mV/m, spanning the range from $\sim$0.02 mV/m (the method's sensitivity limit) to 2 mV/m, was obtained by analysis of particle and magnetic-field data from the Apollo 15 and 16 subsatellites [109]. The analysis also shows that the sense of the electric field is almost always in the direction from dawn to dusk, but it is often variable both in time and space. Also, the plasma sheet characteristics at lunar distance and its interactions with the lunar surface were investigated from electron measurements by Charged Particle Lunar Environment Experiment (CPLEE) deployed on the lunar surface [140–143] and by Explore 35 [124]. Apollo 15 subsatellite magnetometer measurements suggested the existence, on average, of a dayside plasma depletion in the plasma sheet as a result of interactions between the Moon and predominantly earthward, though variable, flow in the plasma sheet [160]. At present, the global Moon-plasma interaction in the Earth's magnetotail remains poorly understood. Reexamination of the extended Moon-plasma interaction region utilizing the plasma and field data of greatly improved quality from the modern instruments has just begun [63, 71, 150].

1.3.5 Lunar Surface Charging

The electron VDF observations by Lunar Prospector on the lunar night side revealed energy-dependent loss cones, indicating electron reflection by both crustal magnetic fields and downward electric fields above the negatively-charged nightside surface [68]. Magnitude of the negative potentials on the nightside lunar surface depends on the ambient plasma conditions, especially directly on the incident electron temperature [64, 150]. The secondary electron yield determined from nightside Lunar Prospector measurements is a factor of $\sim$3 lower than measured for samples in the laboratory [65]. During solar energetic particle (SEP) events, negative surface potentials on the nightside reach up to $-4.5\,\mathrm{kV}$ [62, 66]. For the most part, the nightside lunar surface potentials can be explained by a current balance model between the currents from the ambient plasma and secondary electrons.

Meanwhile, dayside lunar surface potentials appear to be not as straightforward to interpret as the nightside potentials. Surprisingly, it turned out that negative potentials can exist above the dayside lunar surface in the presence of hot electrons of the Earth's magnetotail [63, 67]. These negative dayside potentials cannot be explained by a simple current balance model, which only predicts slightly positive dayside potentials due to the dominant photoelectron current. Instead, a "non-monotonic potential structure" model [44, 57, 129] was applied to the lunar dayside and numerical simulations successfully reproduced the observed negative potentials [132, 133, 136].

As mentioned in Sect. 1.2.5, the photoelectron energy distribution dominantly controls the sunlit surface potentials in the extremely low-density environment in the terrestrial magnetotail lobes. It is most likely that spacecraft photoelectrons have high-energy tail populations, since dozens of spacecraft have observed large positive potentials in the tail lobe. On the other hand, the energy distribution of photoelectrons originating from the lunar surface remains unclear. When the Moon was located in the tail lobes, CPLEE detected lunar photoelectrons that returned to the surface with energies of up to $200\,\mathrm{eV}$ [139]. The observed electron energy distribution and the inferred large positive potential of $\sim +200\,\mathrm{V}$ were explained by assuming a constant photoelectron yield function for the high-energy tail of solar photons. However, this assumption is inconsistent with the experimental result from the returned lunar fines, which showed that the photoelectron yield drops off steeply for higher-energy incident photons after a peak at $14\,\mathrm{eV}$ [40]. The calculated energy distribution under solar irradiation exhibits a mean energy of $2.2\,\mathrm{eV}$ and an extremely weak tail. At present, it is uncertain whether or not the photoelectron yield of the lunar surface remains high for high-energy incident solar photons.

Dayside and terminator surface potentials in the solar wind were inferred from the SIDE observations on the lunar surface [39, 42, 102]. They derived dayside potentials of $\sim +10\,\mathrm{V}$ at solar zenith angles (SZAs) of 20–45°, and near-terminator potentials from ~ -10 to $-100\,\mathrm{V}$ at SZAs of 60–90°. Since the positive potential events seemed to be observed in the Earth's foreshock regions, it was proposed that the somewhat large positive potentials of $\sim +10\,\mathrm{V}$ were related to enhanced secondary electron emission caused by foreshock electron beams [24]. More recently, dayside and near-terminator surface potentials were investigated for various solar-wind parameters using a numerical model, indicating that the dayside potentials vary significantly depending on the input photoelectron current and temperature [165]. Information about the photoelectron energy distribution is crucial for determining how large the lunar dayside surface potentials would be in the tail lobes and the solar wind.

Electric fields arising from the charged lunar surface can lift charged dust grains, which potentially cause so-called "streamers" (sunlight scattered at the terminator) observed by surface landers and astronauts in orbit [53, 107, 108, 113, 129, 133, 166, 167]. Along those lines, I note that dayside and near-terminator surface potential distributions can be very complex due to local surface topography such as craters, which may cause significant dust transportation [11, 33, 35, 36, 136, 194, 195].

1.4 Lunar Exosphere

It is often stated that the Moon has no atmosphere, but it does have an "atmosphere" which is in fact a tenuous, collisionless, and surface-bounded exosphere. Since the mean free path is much greater than the atmospheric scale hight, an exospheric neutral atom or molecule rarely collides with another particle along its ballistic trajectory. The lunar exosphere and its resulting ionosphere are too thin to form an induced magnetosphere and to stand off the solar wind [5, 87]. In this sense, plasma of lunar origin does not act as an effective obstacle to the solar wind and is usually thought to be negligible in terms of the global Moon-solar wind interaction. Some basic information about the lunar exosphere is introduced in this section.

1.4.1 Species and Abundances

The lunar exosphere has been investigated from Apollo surface measurements, ground- and space-based spectroscopy, and pickup ion observations [164]. The measured species include the alkali atoms Na and K, the noble gases He, Ar and Rn, nitrogen- and carbon-based molecules, [54, 55, 79, 80, 138], and pickup ions H^+, H_2^+, He^{++}, He^+, C^+, O^+, Na^+, Al^+, Si^+, K^+ and Ar^+ [10, 19, 78, 105, 168, 180, 191, 193]. Only upper limits were put on densities of other species from the ground- and space-based spectroscopy. Recent observations of lunar pickup ions in both the solar wind and the Earth's magnetotail further constrained the upper limits on densities of some species [70, 135, 156]. The most abundant species among known ones is Ar, with maximum number density of $\sim 10^5\,\mathrm{cm}^{-3}$ near the surface. Although near-surface densities of the alkali metals Na and K are rather small ($70\,\mathrm{cm}^{-3}$ for Na and $17\,\mathrm{cm}^{-3}$ for K), they can be significant sources of plasma of lunar origin owing to their high photo-ionization rates and large scale heights [69].

1.4.2 Source and Loss Mechanisms

The proposed sources of the lunar exospheric neutrals can be categorized into five groups: (i) thermal, (ii) sputtering, (iii) chemical, (iv) meteoritic, and (v) interior release [164]. Thermal desorption involves sublimation from the uppermost regolith and produces thermal Maxwellian distributions with the surface temperature (<400 K ~ 0.03 eV). Sputtering produces somewhat higher-energy neutrals (~ 0.1 eV for photon sputtering and ~ 1–10 eV for charged-particle sputtering). When a chemical reaction on the uppermost layer of the surface has sufficient excess energy, an atom or molecule is emitted from the surface by "chemical sputtering". Micrometeoroid bombardment produces a cloud of impact generated vapor represented by Maxwellian distributions with characteristic temperatures of 2000–5000 K (0.17–0.43 eV) as well as outgassing of hot or molten surface material (<3000 K). Finally, radiogenic products Ar and Rn can be released from the Moon's interior.

Main loss mechanisms, also called "sinks", include (i) gravitational escape, (ii) ionization loss, (iii) chemical loss, and (iv) condensation [164]. Any atom or molecule emitted with radial velocities greater than the escape speed from the Moon (2.38 km/s) will be directly lost to space due to the lack of collisions. Surface-bounded atoms and molecules in the exosphere may suffer solar UV photoionization, charge exchange or impact ionization. When an exospheric atom or molecule collides with the surface, chemical reactions can lead to bounding of the particle to the surface before being ballistically ejected again. Condensation occurs when neutrals contact the cold nightside surface and become absorbed in the solid surface. Among these loss mechanisms, photoionization is usually the dominant (fastest) process in the near-lunar environment.

1.4.3 Lunar Pickup Ions

Ion pickup is a very common process in the solar system and interstellar space [21]. Once exospheric neutral atoms or molecules become ionized or secondary ions are sputtered from the surface, they start to feel the electric and magnetic field in the ambient plasma. Since the electric and magnetic forces usually dominate the gravitational force, the newly ionized particles are no longer bounded by the surface. In many cases, the perpendicular flow speed is much greater than the initial speeds of the ions, and they follow cycloid trajectories. Gyroradii of the heavy ions are often greater than the lunar radius and these pickup ions are detected in their first gyroorbit by lunar-orbiting spacecraft both in the solar wind [69, 70, 180, 191] and the terrestrial magnetotail lobes [135, 137, 168, 193]. These near-lunar pickup ions have phase-bunched, non-gyrotropic velocity distributions, which are highly unstable and will diffuse in the phase space via wave-particle interactions [18, 174].

1.5 Objectives of the Thesis

The Moon-plasma interactions are characterized by the absence of both a substantial atmosphere and a global intrinsic magnetic field, including direct interaction processes of the solid lunar surface with the ambient plasma such as surface absorption, neutralization, scattering, sputtering, and magnetic and electric reflection of charged particles. Since the Apollo era, an enormous amount of work has addressed the near-lunar plasma environment, which involves a wide variety of fascinating plasma physics and planetary sciences. In particular, the solar wind interactions with the Moon and the lunar crustal magnetic anomalies have attracted the greatest deal of interests. A number of observations indicated the importance of kinetic effects of plasma in many places around the Moon such as in the lunar wake and above the magnetic anomalies.

The Moon provides us with a test bed for the study of fundamental plasma processes which will also work at other bodies in our solar system [73]. For example, knowledge on plasma interactions with the lunar surface will be directly applicable to asteroids, airless moons of other planets, and distant comets. Kinetic processes similar to those working at the lunar magnetic anomalies will also operate in the vicinity of relatively small-scale magnetic fields such as intrinsic magnetic fields of Mercury and Ganymede as well as crustal fields of Mars. Thus, the Moon, which is the closest body to the Earth, represents a basic plasma physics laboratory to investigate plasma interactions with airless bodies throughout the solar system and beyond. With the aim of establishing a fundamental framework of airless-body plasma physics, this thesis provides key observational information of Moon-plasma interactions with a particular emphasis on those in the Earth's magnetotail plasma regime.

This thesis presents a series of new observations from recent lunar missions (i.e., Kaguya, ARTEMIS, and Chandrayaan-1), focusing on the data obtained within the Earth's magnetotail. These near-lunar observations in the Earth's magnetotail have extended our knowledge on the Moon-solar wind interactions to an entirely different plasma regime and revealed new aspects of the physical processes working in the lunar environment. The comparison between the plasma characteristics in the solar wind and those in the terrestrial magnetospheric plasma will help us understand the Moon-plasma interactions from more general and comprehensive viewpoints. Chapter 2 contains Kaguya observations and theoretical considerations of non-gyrotropic electron VDFs near the lunar surface. In Chap. 3, I investigate Moon-related plasma features in the Earth's magnetotail from two-point ARTEMIS observations. Chapter 4 presents the lunar ENA observations by Chandrayaan-1 in the Earth's magnetotail and discusses the different characteristics of the magnetic-anomaly interactions with the terrestrial plasma sheet and the solar wind.

References

1. Allegrini, F., Dayeh, M., Desai, M., Funsten, H., Fuselier, S., Janzen, P., McComas, D., Moüibius, E., Reisenfeld, D., M., D.R., Schwadron, N., Wurz, P.: Lunar energetic neutral atom (ENA) spectra measured by the interstellar boundary explorer (IBEX). Planet. Space Sci. **85**(0), 232–242 (2013). doi:10.1016/j.pss.2013.06.014. http://www.sciencedirect.com/science/article/pii/S0032063313001554
2. Anderson, B.J., Acuña, M.H., Korth, H., Purucker, M.E., Johnson, C.L., Slavin, J.A., Solomon, S.C., McNutt, R.L.: The structure of mercury's magnetic field from MESSENGER's first flyby. Science **321**(5885), 82–85 (2008). doi:10.1126/science.1159081. http://www.sciencemag.org/content/321/5885/82.abstract
3. Anderson, B.J., Johnson, C.L., Korth, H., Purucker, M.E., Winslow, R.M., Slavin, J.A., Solomon, S.C., McNutt Jr, R.L., Raines, J.M., Zurbuchen, T.H.: The global magnetic field

of mercury from MESSENGER orbital observations. Science **333**(6051), 1859–1862 (2011). doi:10.1126/science.1211001
4. Anderson, K.A.: Method to determine sense and magnitude of electric field from lunar particle shadows. J. Geophys. Res. **75**, 2591–2594 (1970). doi:10.1029/JA075i013p02591
5. Ando, H., Imamura, T., Nabatov, A., Futaana, Y., Iwata, T., Hanada, H., Matsumoto, K., Mochizuki, N., Kono, Y., Noda, H., Liu, Q., Oyama, K.I., Yamamoto, Z., Saito, A.: Dual-spacecraft radio occultation measurement of the electron density near the lunar surface by the SELENE mission. J. Geophys. Res. **117**, A08313 (2012). doi:10.1029/2011JA017141
6. Backes, H., Neubauer, F., Dougherty, M., Achilleos, N., Andre, N., Arridge, C., Bertucci, C., Jones, G., Khurana, K., Russell, C., Wennmacher, A.: Titan's magnetic field signature during the first Cassini encounter. Science **308**(5724), 992–995 (2005). doi:10.1126/science.1109763
7. Bale, S., Owen, C., Bougeret, J., Goetz, K., Kellogg, P., Lepping, R., Manning, R., Monson, S.: Evidence of currents and unstable particle distributions in an extended region around the lunar plasma wake. Geophys. Res. Lett. **24**(11), 1427–1430 (1997). doi:10.1029/97GL01193
8. Bamford, R.A., Kellett, B., Bradford, W.J., Norberg, C., Thornton, A., Gibson, K.J., Crawford, I.A., Silva, L., Gargaté, L., Bingham, R.: Minimagnetospheres above the lunar surface and the formation of lunar swirls. Phys. Rev. Lett. **109**(081), 101 (2012). doi:10.1103/PhysRevLett.109.081101. http://link.aps.org/doi/10.1103/PhysRevLett.109.081101
9. Barabash, S.: Classes of the solar wind interactions in the solar system. Earth Planets Space **64**, 57–59 (2012)
10. Benson, J., Freeman, J.W., Hills, H.K.: The lunar terminator ionosphere. In: Proceedings of Lunar Science Conference, 6th, pp. 3013–3021 (1975)
11. Berg, O.E.: A Lunar Terminator Configuration. Earth Planet. Sci. **39**, 377–381 (1978)
12. Birch, P.C., Chapman, S.C.: Particle-in-cell simulations of the lunar wake with high phase space resolution. Geophys. Res. Lett. **28**(2), 219–222 (2001). doi:10.1029/2000GL011958. http://dx.doi.org/10.1029/2000GL011958
13. Blewett, D.T., Coman, E.I., Hawke, B.R., Gillis-Davis, J.J., Purucker, M.E., Hughes, C.G.: Lunar swirls: Examining crustal magnetic anomalies and space weathering trends. J. Geophys. Res. **116**, E02002 (2011). doi:10.1029/2010JE003656
14. Borisov, N., Mall, U.: Plasma distribution and electric fields behind the Moon. Phys. Lett. A **265**(5–6), 369–376 (2000). doi:10.1016/S0375-9601(99)00852-X. http://www.sciencedirect.com/science/article/pii/S037596019900852X
15. Borisov, N., Mall, U.: The structure of the double layer behind the Moon. J. Plasma Phys. **67**, 277–299 (2002). doi:10.1017/S0022377802001654
16. Cahill, L., Patel, V.: The boundary of the geomagnetic field, august to november, 1961. Planet. Space Sci. **15**(6), 997–1033 (1967). doi:10.1016/0032-0633(67)90168-7. http://www.sciencedirect.com/science/article/pii/0032063367901687
17. Chapman, S., Ferraro, V.: A new theory of magnetic storms. Nature **126**, 129–130 (1930). doi:10.1038/126129a0
18. Chi, P., Russell, C., Wei, H., Farrell, W.: Observations of narrowband ion cyclotron waves on the surface of the moon in the terrestrial magnetotail. Planet. Space Sci. (2013). doi:10.1016/j.pss.2013.08.020. http://www.sciencedirect.com/science/article/pii/S00320633130%02250
19. Cladis, J.B., Francis, W.E., Vondrak, R.R.: Transport toward earth of ions sputtered from the moon's surface by the solar wind. J. Geophys. Res. **99**(A1), 53–64 (1994). doi:10.1029/93JA02672. http://dx.doi.org/10.1029/93JA02672
20. Clay, D.R., Goldstein, B.E., Neugebauer, M., Snyder, C.W.: Lunar surface solar wind observations at the Apollo 12 and Apollo 15 site. J. Geophys. Res. **80**, 1751–1760 (1975). doi:10.1029/JA080i013p01751
21. Coates, A.J., Jones, G.H.: Plasma environment of Jupiter family comets. Planet. Space Sci. **57**(10, Sp. Iss. SI), 1175–1191 (2009). doi:10.1016/j.pss.2009.04.009
22. Colburn, D.S., Currie, R.G., Mihalov, J.D., Sonett, C.P.: Diamagnetic solar-wind cavity discovered behind Moon. Science **158**, 1040–1042 (1967). doi:10.1126/science.158.3804.1040

23. Colburn, D.S., Mihalov, J.D., Sonett, C.P.: Magnetic observations of the lunar cavity. J. Geophys. Res. **76**, 2940–2957 (1971). doi:10.1029/JA076i013p02940
24. Collier, M.R., Hills, H.K., Stubbs, T.J., Halekas, J.S., Delory, G.T., Espley, J., Farrell, W.M., Freeman, J.W., Vondrak, R.: Lunar surface electric potential changes associated with traversals through the earth's foreshock. Planet. Space Sci. **59**(14), 1727–1743 (2011). doi:10.1016/j.pss.2010.12.010. http://www.sciencedirect.com/science/article/pii/S00320633100%03703
25. DeForest, S.E.: Spacecraft charging at synchronous orbit. J. Geophys. Res. **77**(4), 651–659 (1972). doi:10.1029/JA077i004p00651. http://dx.doi.org/10.1029/JA077i004p00651
26. Dhanya, M.B., Bhardwaj, A., Futaana, Y., Fatemi, S., Holmström, M., Barabash, S., Wieser, M., Wurz, P., Alok, A., Thampi, R.S.: Proton entry into the near-lunar plasma wake for magnetic field aligned flow. Geophys. Res. Lett. **40**(12), 2913–2917 (2013). doi:10.1002/grl.50617. http://dx.doi.org/10.1002/grl.50617
27. Dubinin, E., Fränz, M., Woch, J., Roussos, E., Barabash, S., Lundin, R., Winningham, J.D., Frahm, R.A., Acuña, M.: Plasma morphology at mars. Aspera-3 observations. Space Sci. Rev. **126**(1–4), 209–238 (2006). doi:10.1007/s11214-006-9039-4, Workshop on on Solar Wind Interaction and Atmosphere Evolution of Mars, Kiruna, SWEDEN, 27 Feb–01 Mar 2006
28. Dyal, P., Parkin, C.W., Daily, W.D.: Magnetism and the interior of the Moon. Rev. Geophys. Space Phys. **12**, 568–591 (1974). doi:10.1029/RG012i004p00568
29. Dyal, P., Parkin, C.W., Snyder, C.W., Clay, D.R.: Measurements of lunar magnetic field interaction with the solar wind. Nature **236**(5347), 381–385 (1972). doi:10.1038/236381a0
30. Elphic, R.C., Funsten, H.O., Barraclough, B.L., McComas, D.J., Paffett, M.T., Vaniman, D.T., Heiken, G.: Lunar surface composition and solar wind-induced secondary ion mass spectrometry. Geophys. Res. Lett. **18**(11), 2165–2168 (1991). doi:10.1029/91GL02669. http://dx.doi.org/10.1029/91GL02669
31. Farrell, W., Fitzenreiter, R., Owen, C., Byrnes, J., Lepping, R., Ogilvie, K., Neubauer, F.: Upstream ULF waves and energetic electrons associated with the lunar wake: detection of precursor activity. Geophys. Res. Lett. **23**(10), 1271–1274 (1996). doi:10.1029/96GL01355
32. Farrell, W., Kaiser, M., Steinberg, J., Bale, S.: A simple simulation of a plasma void: Applications to wind observations of the lunar wake. J. Geophys. Res. **103**(A10), 23653–23660 (1998). doi:10.1029/97JA03717
33. Farrell, W.M., Stubbs, T.J., Delory, G.T., Vondrak, R.R., Collier, M.R., Halekas, J.S., Lin, R.P.: Concerning the dissipation of electrically charged objects in the shadowed lunar polar regions. Geophys. Res. Lett. **35**, L19104 (2008). doi:10.1029/2008GL034785
34. Farrell, W.M., Stubbs, T.J., Halekas, J.S., Delory, G.T., Collier, M.R., Vondrak, R.R., Lin, R.P.: Loss of solar wind plasma neutrality and affect on surface potentials near the lunar terminator and shadowed polar regions. Geophys. Res. Lett. **35**(5), L05105 (2008). doi:10.1029/2007GL032653
35. Farrell, W.M., Stubbs, T.J., Halekas, J.S., Killen, R.M., Delory, G.T., Collier, M.R., Vondrak, R.R.: Anticipated electrical environment within permanently shadowed lunar craters. J. Geophys. Res. **115**, E03004 (2010). doi:10.1029/2009JE003464
36. Farrell, W.M., Stubbs, T.J., Vondrak, R.R., Delory, G.T., Halekas, J.S.: Complex Electric Fields Near the Lunar Terminator: The Near-Surface Wake and Accelerated Dust. Geophys. Res. Lett. **34**, L14201 (2007). doi:10.1029/2007GL029312
37. Fatemi, S., Holmström, M., Futaana, Y.: The effects of lunar surface plasma absorption and solar wind temperature anisotropies on the solar wind proton velocity space distributions in the low-altitude lunar plasma wake. J. Geophys. Res. **117**, A10105 (2012). doi:10.1029/2011JA017353
38. Fatemi, S., Holmström, M., Futaana, Y., Barabash, S., Lue, C.: The lunar wake current systems. Geophys. Res. Lett. **40**, 17–21 (2013). doi:10.1029/2012GL054635. http://dx.doi.org/10.1029/2012GL054635
39. Fenner, M.A., Freeman Jr., J.W., Hills, H.K.: The electric potential of the lunar surface. In: Proceedings of lunar science conference, 4th, p. 2877 (1973)

40. Feuerbacher, B., Anderegg, M., Fitton, B., Laude, L.D., Willis, R.F., Grard, R.J.L.: Photoemission from lunar surface fines and the lunar photoelectron sheath. Geochim. Cosmochim. Acta. **3**, 2655–2663 (1972)
41. Feuerbacher, B., Fitton, B.: Experimental investigation of photoemission from satellite surface materials. J. Appl. Phys. **43**(4), 1563–1572 (1972). doi:10.1063/1.1661362. http://link.aip.org/link/?JAP/43/1563/1
42. Freeman, J., Fenner, M., Hills, H.: Electric potential of the moon in the solar wind. J. Geophys. Res. **78**(22), 4560–4567 (1973). doi:10.1029/JA078i022p04560
43. Freeman Jr, J.W.: Energetic ion bursts on the nightside of the Moon. J. Geophys. Res. **77**(1), 239–243 (1972). doi:10.1029/JA077i001p00239
44. Fu, J.H.M.: Surface potential of a photoemitting plate. J. Geophys. Res. **76**(10), 2506–2509 (1971). doi:10.1029/JA076i010p02506
45. Fuller, M.: Lunar magnetism. Rev. Geophys. Space Phys. **12**(1), 23–70 (1974). doi:10.1029/RG012i001p00023
46. Funsten, H.O., Allegrini, F., Bochsler, P.A., Fuselier, S.A., Gruntman, M., Henderson, K., Janzen, P.H., Johnson, R.E., Larsen, B.A., Lawrence, D.J., McComas, D.J., Möbius, E., Reisenfeld, D.B., Rodríguez, D., Schwadron, N.A., Wurz, P.: Reflection of solar wind hydrogen from the lunar surface. J. Geophys. Res. **118**(2), 292–305 (2013). doi:10.1002/jgre.20055. http://dx.doi.org/10.1002/jgre.20055
47. Futaana, Y., Barabash, S., Holmström, M., Bhardwaj, A.: Low energy neutral atoms imaging of the Moon. Planet. Space Sci. **54**(2), 132–143 (2006). doi:10.1016/j.pss.2005.10.010
48. Futaana, Y., Barabash, S., Wieser, M., Holmström, M., Bhardwaj, A., Dhanya, M.B., Sridharan, R., Wurz, P., Schaufelberger, A., Asamura, K.: Protons in the near-lunar wake observed by the Sub-keV Atom Reflection Analyzer on board Chandrayaan-1. J. Geophys. Res. **115**, A10248 (2010). doi:10.1029/2010JA015264
49. Futaana, Y., Barabash, S., Wieser, M., Holmström, M., Lue, C., Wurz, P., Schaufelberger, A., Bhardwaj, A., Dhanya, M.B., Asamura, K.: Empirical energy spectra of neutralized solar wind protons from the lunar regolith. J. Geophys. Res. **117**, E05005 (2012). doi:10.1029/2011JE004019
50. Futaana, Y., Barabash, S., Wieser, M., Lue, C., Wurz, P., Vorburger, A., Bhardwaj, A., Asamura, K.: Remote energetic neutral atom imaging of electric potential over a lunar magnetic anomaly. Geophys. Res. Lett. **40**, 262–266 (2013). doi:10.1002/grl.50135. http://dx.doi.org/10.1002/grl.50135
51. Futaana, Y., Machida, S., Saito, Y., Matsuoka, A., Hayakawa, H.: Counterstreaming electrons in the near vicinity of the Moon observed by plasma instruments on board NOZOMI. J. Geophys. Res. **106**, 18729–18740 (2001). doi:10.1029/2000JA000146
52. Garrick-Bethell, I., Head III, J.W., Pieters, C.M.: Spectral properties, magnetic fields, and dust transport at lunar swirls. Icarus **212**(2), 480–492 (2011). doi:10.1016/j.icarus.2010.11.036
53. Goertz, C.K.: Dusty plasmas in the solar system. Rev. Geophys. **27**(2), 271–292 (1989). doi:10.1029/RG027i002p00271
54. Gorenstein, P., Bjorkholm, P.: Detection of radon emanation from the crater aristarchus by the apollo 15 alpha particle spectrometer. Science **179**(4075), 792–794 (1973). doi:10.1126/science.179.4075.792. http://www.sciencemag.org/content/179/4075/792.abstract
55. Gorenstein, P., Golub, L., Bjorkholm, P.: Radon emanation from the moon, spatial and temporal variability. The Moon **9**(1–2), 129–140 (1974). doi:10.1007/BF00565399. http://dx.doi.org/10.1007/BF00565399
56. Grard, R.J.L.: Properties of the satellite photoelectron sheath derived from photoemission laboratory measurements. J. Geophys. Res. **78**(16), 2885–2906 (1973). doi:10.1029/JA078i016p02885. http://dx.doi.org/10.1029/JA078i016p02885
57. Guernsey, R.L., Fu, J.H.M.: Potential distribution surrounding a photo-emitting, plate in a dilute plasma. J. Geophys. Res. **75**(16), 3193–3199 (1970). doi:10.1029/JA075i016p03193
58. Halekas, J.S., Bale, S.D., Mitchell, D.L., Lin, R.P.: Electrons and magnetic fields in the lunar plasma wake. J. Geophys. Res. **110**, A07222 (2005). doi:10.1029/2004JA010991

59. Halekas, J.S., Brain, D.A., Lin, R.P., Mitchell, D.L.: Solar wind interaction with lunar crustal magnetic anomalies. Adv. Space Res. **41**(8), 1319–1324 (2008). doi:10.1016/j.asr.2007.04.003
60. Halekas, J.S., Brain, D.A., Mitchell, D.L., Lin, R.P.: Whistler waves observed near lunar crustal magnetic sources. Geophys. Res. Lett. **33**, L22104 (2006). doi:10.1029/2006GL027684
61. Halekas, J.S., Brain, D.A., Mitchell, D.L., Lin, R.P., Harrison, L.: On the occurrence of magnetic enhancements caused by solar wind interaction with lunar crustal fields. Geophys. Res. Lett. **33**(8), L08106 (2006). doi:10.1029/2006GL025931
62. Halekas, J.S., Delory, G.T., Brain, D.A., Lin, R.P., Fillingim, M.O., Lee, C.O., Mewaldt, R.A., Stubbs, T.J., Farrell, W.M., Hudson, M.K.: Extreme Lunar Surface Charging during Solar Energetic Particle Events. Geophys. Res. Lett. **34**, L02111 (2007). doi:10.1029/2006GL028517
63. Halekas, J.S., Delory, G.T., Farrell, W.M., Angelopoulos, V., McFadden, J.P., Bonnell, J.W., Fillingim, M.O., Plaschke, F.: First Remote Measurements of Lunar Surface Charging from ARTEMIS: Evidence for Nonmonotonic Sheath Potentials Above the Dayside Surface. J. Geophys. Res. **116**, A07103 (2011). doi:10.1029/2011JA016542
64. Halekas, J.S., Delory, G.T., Lin, R.P., Stubbs, T.J., Farrell, W.M.: Lunar Prospector Observations of the Electrostatic Potential of the Lunar Surface and its Response to Incident Currents. J. Geophys. Res. **113**, A09102 (2008). doi:10.1029/2008JA013194
65. Halekas, J.S., Delory, G.T., Lin, R.P., Stubbs, T.J., Farrell, W.M.: Lunar prospector measurements of secondary electron emission from lunar Regolith. Planet. Space Sci. **57**, 78–82 (2009). doi:10.1016/j.pss.2008.11.009
66. Halekas, J.S., Delory, G.T., Lin, R.P., Stubbs, T.J., Farrell, W.M.: Lunar surface charging during solar energetic particle events: Measurement and prediction. J. Geophys. Res. **114**, A05110 (2009). doi:10.1029/2009JA014113
67. Halekas, J.S., Lin, R.P., Mitchell, D.L.: Large negative lunar surface potentials in sunlight and shadow. Geophys. Res. Lett. **32**, L09102 (2005). doi:10.1029/2005GL022627
68. Halekas, J.S., Mitchell, D.L., Lin, R.P., Hood, L.L., Acuña, M.H., Binder, A.B.: Evidence for negative charging of the Lunar surface in shadow. Geophys. Res. Lett. **29**, 1435 (2002). doi:10.1029/2001GL014428
69. Halekas, J.S., Poppe, A.R., Delory, G.T., Sarantos, M., Farrell, W.M., Angelopoulos, V., McFadden, J.P.: Lunar pickup ions observed by artemis: Spatial and temporal distribution and constraints on species and source locations. J. Geophys. Res. **117**, E06006 (2012). doi:10.1029/2012JE004107
70. Halekas, J.S., Poppe, A.R., Delory, G.T., Sarantos, M., McFadden, J.P.: Using artemis pickup ion observations to place constraints on the lunar atmosphere. J. Geophys. Res. **118**, 81–88 (2013). doi:10.1029/2012JE004292. http://dx.doi.org/10.1029/2012JE004292
71. Halekas, J.S., Poppe, A.R., Farrell, W.M., Delory, G.T., Angelopoulos, V., McFadden, J.P., Bonnell, J.W., Glassmeier, K.H., Plaschke, F., Roux, A., Ergun, R.E.: Lunar precursor effects in the solar wind and terrestrial magnetosphere. J. Geophys. Res. **117**, A05101 (2012). doi:10.1029/2011JA017289
72. Halekas, J.S., Poppe, A.R., McFadden, J.P., Glassmeier, K.H.: The effects of reflected protons on the plasma environment of the moon for parallel interplanetary magnetic fields. Geophys. Res. Lett. **40**(17), 4544–4548 (2013). doi:10.1002/grl.50892. http://dx.doi.org/10.1002/grl.50892
73. Halekas, J.S., Saito, Y., Delory, G.T., Farrell, W.M.: New views of the lunar plasma environment. Planet. Space Sci. **59**(14, SI), 1681–1694 (2011). doi:10.1016/j.pss.2010.08.011. First Workshop on Lunar Dust, Plasma and Atmosphere—The Next Steps (LDAP), Boulder, CO, 27–29 Jan 2010
74. Harnett, E., Winglee, R.: Two-dimensional MHD simulation of the solar wind interaction with magnetic field anomalies on the surface of the Moon. J. Geophys. Res. **105**(A11), 24997–25007 (2000). doi:10.1029/2000JA000074

75. Harnett, E., Winglee, R.: 2.5D Particle and MHD simulations of mini-magnetospheres at the Moon. J. Geophys. Res. **107**(A12), 1421 (2002). doi:10.1029/2002JA009241
76. Hashimoto, K., Hashitani, M., Kasahara, Y., Omura, Y., Nishino, M.N., Saito, Y., Yokota, S., Ono, T., Tsunakawa, H., Shibuya, H., Matsushima, M., Shimizu, H., Takahashi, F.: Electrostatic solitary waves associated with magnetic anomalies and wake boundary of the Moon observed by KAGUYA. Geophys. Res. Lett. **37**, L19204 (2010). doi:10.1029/2010GL044529
77. Hemingway, D., Garrick-Bethell, I.: Magnetic field direction and Lunar swirl morphology: insights from Airy and Reiner Gamma. J. Geophys. Res. **117**, E10012 (2012). doi:10.1029/2012JE004165
78. Hilchenbach, M., Hovstadt, D., Klecker, B., Möbius, E.: Observation of energetic lunar pick-up ions near earth. Adv. Space Res. **13**(10), 321–324 (1993). doi:10.1016/0273-1177(93)90086-Q
79. Hoffman, J., Hodges R.R., J.: Molecular gas species in the lunar atmosphere. The Moon **14**(1), 159–167 (1975). doi:10.1007/BF00562981. http://dx.doi.org/10.1007/BF00562981
80. Hoffman, J.H., Hodges Jr., R.R., Johnson, F.S., Evans, D.E.: Lunar atmospheric composition results from Apollo 17. Proceedings of Lunar Science Conference, 4th, pp. 2865 (1973)
81. Holmström, M., Fatemi, S., Futaana, Y., Nilsson, H.: The interaction between the moon and the solar wind. Earth Planets Space **64**, 237–245 (2012)
82. Holmström, M., Wieser, M., Barabash, S., Futaana, Y., Bhardwaj, A.: Dynamics of solar wind protons reflected by the Moon. J. Geophys. Res. **115**, A06206 (2010). doi:10.1029/2009JA014843
83. Hood, L.L., Williams, C.R.: The lunar swirls—distribution and possible origins. Proceedings of Lunar Planetery Science Conference, 19th, pp. 99–113 (1989)
84. Hood, L.L., Zakharian, A., Halekas, J.S., Mitchell, D.L., Lin, R.P., Acuña, M.H., Binder, A.B.: Initial mapping and interpretation of lunar crustal magnetic anomalies using lunar prospector magnetometer data. J. Geophys. Res. **106**, 27825–27839 (2001). doi:10.1029/2000JE001366
85. Hutchinson, I.H.: Electron velocity distribution instability in magnetized plasma wakes and artificial electron mass. J. Geophys. Res. **117**, A03101 (2012). doi:10.1029/2011JA017119
86. Hutchinson, I.H.: Near-lunar proton velocity distribution explained by electrostatic acceleration. J. Geophys. Res. **118**(5), 1825–1827 (2013). doi:10.1002/jgra.50277. http://dx.doi.org/10.1002/jgra.50277
87. Imamura, T., Nabatov, A., Mochizuki, N., Iwata, T., Hanada, H., Matsumoto, K., Noda, H., Kono, Y., Liu, Q., Futaana, Y., Ando, H., Yamamoto, Z., Oyama, K.I., Saito, A.: Radio occultation measurement of the electron density near the lunar surface using a subsatellite on the selene mission. J. Geophys. Res. **117**, A06303 (2012). doi:10.1029/2011JA017293
88. Israelevich, P., Ofman, L.: Hybrid simulation of the shock wave trailing the Moon. J. Geophys. Res. **117**, A08223 (2012). doi:10.1029/2011JA017358
89. Johnson, C.L., Purucker, M.E., Korth, H., Anderson, B.J., Winslow, R.M., Asad, M.M.H.A., Slavin, J.A., Alexeev, I.I., Phillips, R.J., Zuber, M.T., Solomon, S.C.: Messenger observations of mercury's magnetic field structure. J. Geophys. Res. **117**, E00L14 (2012). doi:10.1029/2012JE004217
90. Johnson, F.S., Midgley, J.E.: Notes on the lunar magnetosphere. J. Geophys. Res. **73**(5), 1523–000 (1968). doi:10.1029/JA073i005p01523
91. Kallio, E.: Formation of the lunar wake in quasi-neutral hybrid model. Geophys. Res. Lett. **32**(6), L06107 (2005). doi:10.1029/2004GL021989. http://dx.doi.org/10.1029/2004GL021989
92. Kallio, E., Jarvinen, R., Dyadechkin, S., Wurz, P., Barabash, S., Alvarez, F., Fernandes, V.A., Futaana, Y., Harri, A.M., Heilimo, J., Lue, C., Mäkelä, J., Porjo, N., Schmidt, W., Siili, T.: Kinetic simulations of finite gyroradius effects in the lunar plasma environment on global, meso, and microscales. Planet. Space Sci. **74**, 146–155 (2012). doi:10.1016/j.pss.2012.09.012
93. Kallio, E., Sillanpää, I., Janhunen, P.: Titan in subsonic and supersonic flow. Geophys. Res. Lett. **31**(15), L15703 (2004). doi:10.1029/2004GL020344. http://dx.doi.org/10.1029/2004GL020344

94. Katz, I., Parks, D.E., Mandell, M.J., Harvey, J.M., Brownell Jr., D.H., Wang, S.S., Rotenberg, M.: A three dimensional dynamic study of electrostatic charging in materials. Technical report (1977)
95. Khurana, K.K., Russell, C.T., Dougherty, M.K.: Magnetic portraits of Tethys and Rhea. Icarus **193**(2), 465–474 (2008). doi:10.1016/j.icarus.2007.08.005
96. Kimura, S., Nakagawa, T.: Electromagnetic full particle simulation of the electric field structure around the Moon and the Lunar Wake. Earth Planets Space **60**, 591–599 (2008)
97. Kivelson, M., Khurana, K., Coroniti, F., Joy, S., Russell, C., Walker, R., Warnecke, J., Bennett, L., Polanskey, C.: The magnetic field and magnetosphere of Ganymede. Geophys. Res. Lett. **24**(17), 2155–2158 (1997). doi:10.1029/97GL02201
98. Kivelson, M., Khurana, K., Russell, C., Walker, R., Warnecke, J., Coroniti, F., Polanskey, C., Southwood, D., Schubert, G.: Discovery of Ganymede's magnetic field by the Galileo spacecraft. Nature **384**(6609), 537–541 (1996). doi:10.1038/384537a0
99. Kurata, M., Tsunakawa, H., Saito, Y., Shibuya, H., Matsushima, M., Shimizu, H.: Mini-magnetosphere over the Reiner Gamma magnetic anomaly region on the Moon. Geophys. Res. Lett. **32**(24), L24205 (2005). doi:10.1029/2005GL024097
100. Leblanc, F., Doressoundiram, A., Schneider, N., Mangano, V., Ariste, A.L., Lemen, C., Gelly, B., Barbieri, C., Cremonese, G.: High latitude peaks in Mercury's sodium exosphere: spectral signature using THEMIS solar telescope. Geophys. Res. Lett. **35**(18), L18204 (2008). doi:10.1029/2008GL035322
101. Lin, R.P., Mitchell, D.L., Curtis, D.W., Anderson, K.A., Carlson, C.W., McFadden, J., Acuña, M.H., Hood, L.L., Binder, A.: Lunar surface magnetic fields and their interaction with the solar wind: results from Lunar prospector. Science **281**, 1480–1484 (1998). doi:10.1126/science.281.5382.1480
102. Lindeman, R., Freeman Jr., J.W., Vondrak, R.R.: Ions from the lunar atmosphere. Proceedings of Lunar Science Conference, 4th, pp. 2889 (1973)
103. Lue, C., Futaana, Y., Barabash, S., Wieser, M., Holmström, M., Bhardwaj, A., Dhanya, M.B., Wurz, P.: Strong influence of lunar crustal fields on the solar wind flow. Geophys. Res. Lett. **38**, L03202 (2011). doi:10.1029/2010GL046215
104. Lyon, E.F., Bridge, H.S., Binsack, J.H.: Explorer 35 plasma measurements in the vicinity of the Moon. J. Geophys. Res. **72**, 6113–6117 (1967). doi:10.1029/JZ072i023p06113
105. Mall, U., Kirsch, E., Cierpka, K., Wilken, B., Söding, A., Neubauer, F., Gloeckler, G., Galvin, A.: Direct observation of lunar pick-up ions near the Moon. Geophys. Res. Lett. **25**(20), 3799–3802 (1998). doi:10.1029/1998GL900003
106. McComas, D.J., Allegrini, F., Bochsler, P., Frisch, P., Funsten, H.O., Gruntman, M., Janzen, P.H., Kucharek, H., Moebius, E., Reisenfeld, D.B., Schwadron, N.A.: Lunar backscatter and neutralization of the solar wind: first observations of neutral atoms from the Moon. Geophys. Res. Lett. **36**, L12104 (2009). doi:10.1029/2009GL038794
107. McCoy, J.E.: Photometric studies of light scattering above the lunar terminator from Apollo solar corona photography. In: Proceedings of Lunar Science Conference, 7th, pp. 1087–1112 (1976)
108. McCoy, J.E., Criswell, D.R.: Evidence for a high altitude distribution of lunar dust. In: Proceedings of Lunar Science Conference, 5th, pp. 2991–3005 (1974)
109. McCoy, J.E., Lin, R.P., McGuire, R.E., Chase, L.M., Anderson, K.A.: Magnetotail electric fields observed from Lunar orbit. J. Geophys. Res. **80**, 3217–3224 (1975). doi:10.1029/JA080i022p03217
110. McCracken, G.M.: The behaviour of surfaces under ion bombardment. Rep. Prog. Phys. **38**(2), 241 (1975). doi:10.1088/0034-4885/38/2/002. http://stacks.iop.org/0034-4885/38/i=2/a=002
111. McGrath, M.A., Johnson, R.E., Lanzerotti, L.J.: Sputtering of sodium on the planet Mercury. Nature **323**, 694–696 (1986). doi:10.1038/323694a0
112. Michel, F.C.: Magnetic field structure behind the moon. J. Geophys. Res. **73**, 1533–1542 (1968). doi:10.1029/JA073i005p01533

113. Murphy, D.L., Vondrak, R.R.: Effects of levitated dust on astronomical observations from the Lunar surface. In: Proceedings of Lunar Planetery Science Conference, 24th, pp. 889 (1993)
114. Nagy, A., Winterhalter, D., Sauer, K., Cravens, T., Brecht, S., Mazelle, C., Crider, D., Kallio, E., Zakharov, A., Dubinin, E., Verigin, M., Kotova, G., Axford, W., Bertucci, C., Trotignon, J.: The plasma environment of Mars. Space Sci. Rev. **111**(1–2), 33–114 (2004). doi:10.1023/B:SPAC.0000032718.47512.92
115. Nakagawa, T.: Ion entry into the wake behind a nonmagnetized obstacle in the solar wind: Two-dimensional particle-in-cell simulations. J. Geophys. Res. **118**(5), 1849–1860 (2013). doi:10.1002/jgra.50129. http://dx.doi.org/10.1002/jgra.50129
116. Nakagawa, T., Nakayama, A., Takahashi, F., Tsunakawa, H., Shibuya, H., Shimizu, H., Matsushima, M.: Large-amplitude monochromatic ULF waves detected by Kaguya at the Moon. J. Geophys. Res. **117**, A04101 (2012). doi:10.1029/2011JA017249
117. Nakagawa, T., Takahashi, F., Tsunakawa, H., Shibuya, H., Shimizu, H., Matsushima, M.: Non-monochromatic whistler waves detected by Kaguya on the dayside surface of the moon. Earth Planets Space **63**(1), 37–46 (2011). doi:10.5047/eps.2010.01.005. Thirrd Kaguya Science Meeting on Earth, Planets and Space, Tokyo, Japan, 14–15 Jan 2009
118. Nakagawa, T., Takahashi, Y., Iizima, M.: GEOTAIL observation of upstream ULF waves associated with lunar wake. Earth Planets Space **55**(9), 569–580 (2003)
119. Ness, N.F., Behannon, K.W., Lepping, R.P., Whang, Y.C., Schatten, K.H.: Magnetic field observations near mercury: preliminary results from mariner 10. Science **185**(4146), 151–160 (1974). doi:10.1126/science.185.4146.151. http://www.sciencemag.org/content/185/4146/151.abstract
120. Ness, N.F., Behannon, K.W., Scearce, C.S., Cantara, S.C.: Early results from the magnetic field experiment on lunar explorer 35. J. Geophys. Res. **72**, 5769–5778 (1967). doi:10.1029/JZ072i023p05769
121. Ness, N.F., Behannon, K.W., Taylor, H.E., Whang, Y.C.: Perturbations of the interplanetary magnetic field by the lunar wake. J. Geophys. Res. **73**, 3421–3440 (1968). doi:10.1029/JA073i011p03421
122. Neugebauer, M., Snyder, C.W., Clay, D.R., Goldstein, B.E.: Solar wind observations on the lunar surface with the Apollo-12 ALSEP. Planet. Space Sci. **20**, 1577–1591 (1972). doi:10.1016/0032-0633(72)90184-5
123. Niehus, H., Heiland, W., Taglauer, E.: Low-energy ion scattering at surfaces. Surface Science Reports 17(4–5), 213–303 (1993). doi:10.1016/0167-5729(93)90024-J. http://www.sciencedirect.com/science/article/pii/016757299390%024J
124. Nishida, A., Lyon, E.F.: Plasma sheet at lunar distance: structure and solar-wind dependence. J. Geophys. Res. **77**, 4086–4099 (1972). doi:10.1029/JA077i022p04086
125. Nishino, M.N., Fujimoto, M., Maezawa, K., Saito, Y., Yokota, S., Asamura, K., Tanaka, T., Tsunakawa, H., Matsushima, M., Takahashi, F., Terasawa, T., Shibuya, H., Shimizu, H.: Solar-Wind proton access deep into the near-moon wake. Geophys. Res. Lett. **36**, L16103 (2009). doi:10.1029/2009GL039444
126. Nishino, M.N., Fujimoto, M., Saito, Y., Yokota, S., Kasahara, Y., Omura, Y., Goto, Y., Hashimoto, K., Kumamoto, A., Ono, T., Tsunakawa, H., Matsushima, M., Takahashi, F., Shibuya, H., Shimizu, H., Terasawa, T.: Effect of the solar wind proton entry into the deepest lunar wake. Geophys. Res. Lett. **37**, L12106 (2010). doi:10.1029/2010GL043948
127. Nishino, M.N., Fujimoto, M., Tsunakawa, H., Matsushima, M., Shibuya, H., Shimizu, H., Takahashi, F., Saito, Y., Yokota, S.: Control of lunar external magnetic enhancements by IMF polarity: a case study. Planet. Space Sci. **73**(1), 161–167 (2012). doi:10.1016/j.pss.2012.09.011. http://www.sciencedirect.com/science/article/pii/S00320633120%02863
128. Nishino, M.N., Maezawa, K., Fujimoto, M., Saito, Y., Yokota, S., Asamura, K., Tanaka, T., Tsunakawa, H., Matsushima, M., Takahashi, F., Terasawa, T., Shibuya, H., Shimizu, H.: Pairwise energy gain-loss feature of solar wind protons in the near-Moon wake. Geophys. Res. Lett. **36**, L12108 (2009). doi:10.1029/2009GL039049
129. Nitter, T., Havnes, O., Melandso, F.: Levitation and dynamics of charged dust in the photoelectron sheath above surfaces in space. J. Geophys. Res. **103**(A4), 6605–6620 (1998). doi:10.1029/97JA03523

130. Ogilvie, K.W., Steinberg, J.T., Fitzenreiter, R.J., Owen, C.J., Lazarus, A.J., Farrell, W.M., Torbet, R.B.: Observations of the lunar plasma wake from the WIND spacecraft on December 27, 1994. Geophys. Res. Lett. **23**, 1255–1258 (1996). doi:10.1029/96GL01069
131. Pedersen, A.: Solar wind and magnetosphere plasma diagnostics by spacecraft electrostatic potential measurements. Ann. Geophys. **13**(2), 118–129 (1995). doi:10.1007/s00585-995-0118-8
132. Poppe, A., Halekas, J.S., Horányi, M.: Negative potentials above the day-side lunar surface in the terrestrial plasma sheet: Evidence of non-monotonic potentials. Geophys. Res. Lett. **38**, L02103 (2011). doi:10.1029/2010GL046119
133. Poppe, A., Horányi, M.: Simulations of the photoelectron sheath and dust levitation on the lunar surface. J. Geophys. Res. **115**, A08106 (2010). doi:10.1029/2010JA015286
134. Poppe, A.R., Halekas, J.S., Delory, G.T., Farrell, W.M.: Particle-in-cell simulations of the solar wind interaction with lunar crustal magnetic anomalies: magnetic cusp regions. J. Geophys. Res. **117**, A09105 (2012). doi:10.1029/2012JA017844
135. Poppe, A.R., Halekas, J.S., Samad, R., Sarantos, M., Delory, G.T.: Model-based constraints on the lunar exosphere derived from ARTEMIS pickup ion observations in the terrestrial magnetotail. J. Geophys. Res. **118**(5), 1135–1147 (2013). doi:10.1002/jgre.20090. http://dx.doi.org/10.1002/jgre.20090
136. Poppe, A.R., Piquette, M., Likhanskii, A., Horányi, M.: The effect of surface topography on the lunar photoelectron sheath and electrostatic dust transport. Icarus **221**(1), 135–146 (2012). doi:10.1016/j.icarus.2012.07.018
137. Poppe, A.R., Samad, R., Halekas, J.S., Sarantos, M., Delory, G.T., Farrell, W.M., Angelopoulos, V., McFadden, J.P.: ARTEMIS observations of lunar pick-up ions in the terrestrial magnetotail lobes. Geophys. Res. Lett. **39**, L17104 (2012). doi:10.1029/2012GL052909
138. Potter, A.E., Morgan, T.H.: Discovery of sodium and potassium vapor in the atmosphere of the Moon. Science **241**(4866), 675–680 (1988). doi:10.1126/science.241.4866.675. http://www.sciencemag.org/content/241/4866/675.abstract
139. Reasoner, D.L., Burke, W.J.: Characteristics of the lunar photoelectron layer in the geomagnetic tail. J. Geophys. Res. **77**(34), 6671–000 (1972). doi:10.1029/JA077i034p06671
140. Reiff, P.H.: Magnetic shadowing of charged particles by an extended surface. J. Geophys. Res. **81**(19), 3423–3427 (1976). doi:10.1029/JA081i019p03423
141. Reiff, P.H., Burke, W.J.: Interactions of the plasma sheet with the lunar surface at the apollo 14 site. J. Geophys. Res. **81**(25), 4761–4764 (1976). doi:10.1029/JA081i025p04761
142. Rich, F.J., Reasoner, D.L., Burke, W.J.: Plasma sheet at lunar distance: characteristics and interactions with the lunar surface. J. Geophys. Res. **78**(34), 8097–8112 (1973). doi:10.1029/JA078i034p08097
143. Rich, F.J., Reasoner, D.L., Burke, W.J., Hones, E.W.: Plasma sheet at lunar distance during magnetospheric substorm. J. Geophys. Res. **79**(13), 1981–1984 (1974). doi:10.1029/JA079i013p01981
144. Richmond, N.C., Hood, L.L.: A preliminary global map of the vector lunar crustal magnetic field based on Lunar prospector magnetometer data. J. Geophys. Res. **113**(E2), E02010 (2008). doi:10.1029/2007JE002933
145. Richmond, N.C., Hood, L.L., Halekas, J.S., Mitchell, D.L., Lin, R.P., Acuña, M., Binder, A.B.: Correlation of a strong lunar magnetic anomaly with a high-albedo region of the descartes mountains. Geophys. Res. Lett. **30**, 1395 (2003). doi:10.1029/2003GL016938
146. Rodríguez M., D., Saul, L., Wurz, P., Fuselier, S., Funsten, H., McComas, D., Möbius, E.: IBEX-Lo observations of energetic neutral hydrogen atoms originating from the lunar surface. Planet. Space Sci. **60**(1), 297–303 (2012). doi:10.1016/j.pss.2011.09.009. http://www.sciencedirect.com/science/article/pii/S003206331110%02972
147. Roussos, E., Müller, J., Simon, S., Bößwetter, A., Motschmann, U., Krupp, N., Fränz, M., Woch, J., Khurana, K.K., Dougherty, M.K.: Plasma and fields in the wake of Rhea: 3-D hybrid simulation and comparison with Cassini data. Ann. Geophys. **26**(3), 619–637 (2008). doi:10.5194/angeo-26-619-2008. http://www.ann-geophys.net/26/619/2008/

148. Russell, C.T.: The dynamics of planetary magnetospheres. Planet. Space Sci. **49**, 1005–1030 (2001). doi:10.1016/S0032-0633(01)00017-4
149. Russell, C.T., Lichtenstein, B.R.: On the source of lunar limb compressions. J. Geophys. Res. **80**, 4700–4711 (1975). doi:10.1029/JA080i034p04700
150. Saito, Y., Nishino, M., Yokota, S., Tsunakawa, H., Matsushima, M., Takahashi, F., Shibuya, H., Shimizu, H.: Night side lunar surface potential in the Earth's magnetosphere. Adv. Space Res. (2013). doi:10.1016/j.asr.2013.05.011. http://www.sciencedirect.com/science/article/pii/S027311771130%02664
151. Saito, Y., Nishino, M.N., Fujimoto, M., Yamamoto, T., Yokota, S., Tsunakawa, H., Shibuya, H., Matsushima, M., Shimizu, H., Takahashi, F.: Simultaneous observation of the electron acceleration and ion deceleration over lunar magnetic anomalies. Earth Planets Space **64**, 83–92 (2012)
152. Saito, Y., Yokota, S., Asamura, K., Tanaka, T., Akiba, R., Fujimoto, M., Hasegawa, H., Hayakawa, H., Hirahara, M., Hoshino, M., Machida, S., Mukai, T., Nagai, T., Nagatsuma, T., Nakamura, M., ichiro Oyama, K., Sagawa, E., Sasaki, S., Seki, K., Terasawa, T.: Low-energy charged particle measurement by MAP-PACE onboard SELENE. Earth Planets Space **60**, 375–385 (2008)
153. Saito, Y., Yokota, S., Asamura, K., Tanaka, T., Nishino, M.N., Yamamoto, T., Terakawa, Y., Fujimoto, M., Hasegawa, H., Hayakawa, H., Hirahara, M., Hoshino, M., Machida, S., Mukai, T., Nagai, T., Nagatsuma, T., Nakagawa, T., Nakamura, M., ichiro Oyama, K., Sagawa, E., Sasaki, S., Seki, K., Shinohara, I., Terasawa, T., Tsunakawa, H., Shibuya, H., Matsushima, M., Shimizu, H., Takahashi, F.: In-flight performance and initial results of plasma energy angle and composition experiment (PACE) on SELENE (Kaguya). Space Sci. Rev. **154**, 265–303 (2010). doi:10.1007/s11214-010-9647-x
154. Saito, Y., Yokota, S., Tanaka, T., Asamura, K., Nishino, M.N., Fujimoto, M., Tsunakawa, H., Shibuya, H., Matsushima, M., Shimizu, H., Takahashi, F., Mukai, T., Terasawa, T.: Solar wind proton reflection at the lunar surface: low energy ion measurement by MAP-PACE onboard SELENE (KAGUYA). Geophys. Res. Lett. **35**, L24205 (2008). doi:10.1029/2008GL036077
155. Samir Jr, U., Wright, K.H., Stone, N.H.: The expansion of a plasma into a vacuum: basic phenomena and processes and applications to space plasma physics. Rev. Geophys. Space Phys. **21**(7), 1631–1646 (1983). doi:10.1029/RG021i007p01631
156. Sarantos, M., Hartle, R.E., Killen, R.M., Saito, Y., Slavin, J.A., Glocer, A.: Flux estimates of ions from the lunar exosphere. Geophys. Res. Lett. **39**, L13101 (2012). doi:10.1029/2012GL052001
157. Schaufelberger, A., Wurz, P., Barabash, S., Wieser, M., Futaana, Y., Holmström, M., Bhardwaj, A., Dhanya, M.B., Sridharan, R., Asamura, K.: Scattering function for energetic neutral hydrogen atoms off the lunar surface. Geophys. Res. Lett. **38**, L22202 (2011). doi:10.1029/2011GL049362
158. Schmidt, R., Arends, H., Pedersen, A., Riidenauer, F., Fehringer, M., Narheim, B.T., Svenes, R., Kvernsveen, K., Tsuruda, K., Mukai, T., Hayakawa, H., Nakamura, M.: Results from active spacecraft potential control on the geotail spacecraft. J. Geophys. Res. **100**(A9), 17253–17259 (1995). doi:10.1029/95JA01552
159. Schubert, G., Lichtenstein, B.R.: Observations of Moon–Plasma interactions by orbital and surface experiment. Rev. Geophys. Space Phys. **12**, 592–626 (1974). doi:10.1029/RG012i004p00592
160. Schubert, G., Lichtenstein, B.R., Russell C.T., Coleman Jr, P.J., Smith, B.F., Colburn, D.S., Sonett, C.P.: Lunar dayside plasma sheet depletion: Inference from magnetic observations. Geophys. Res. Lett. **1**, 97–100 (1974). doi:10.1029/GL001i003p00097
161. Sigmund, P.: Theory of sputtering. i. sputtering yield of amorphous and polycrystalline targets. Phys. Rev. **184**, 383–416 (1969). doi:10.1103/PhysRev.184.383. http://link.aps.org/doi/10.1103/PhysRev.184.383
162. Simon, S., Saur, J., Neubauer, F.M., Motschmann, U., Dougherty, M.K.: Plasma wake of Tethys: Hybrid simulations versus Cassini MAG data. Geophys. Res. Lett. 36, L04,108 (2009). doi:10.1029/2008GL036943

163. Sonett, C.P., Mihalov, J.D.: Lunar fossil magnetism and perturbations of the solar wind. J. Geophys. Res. **77**, 588–603 (1972). doi:10.1029/JA077i004p00588
164. Stern, S.: The lunar atmosphere: History, status, current problems, and context. Rev. Geophys. **37**(4), 453–491 (1999). doi:10.1029/1999RG900005
165. Stubbs, T., Farrell, W., Halekas, J., Burchill, J., Collier, M., Zimmerman, M., Vondrak, R., Delory, G., Pfaff, R.: Dependence of lunar surface charging on solar wind plasma conditions and solar irradiation. Planet. Space Sci. (2013). doi:10.1016/j.pss.2013.07.008. http://www.sciencedirect.com/science/article/pii/S003206331300201876
166. Stubbs, T., Vondrak, R., Farrell, W.: A dynamic fountain model for lunar dust. Adv. Space Res. **37**(1), 59–66 (2006). doi:10.1016/j.asr.2005.04.048
167. Stubbs, T.J., Glenar, D.A., Farrell, W.M., Vondrak, R.R., Collier, M.R., Halekas, J.S., Delory, G.T.: On the role of dust in the lunar ionosphere. Planet. Space Sci. **59**(13), 1659–1664 (2011). doi:10.1016/j.pss.2011.05.011
168. Tanaka, T., Saito, Y., Yokota, S., Asamura, K., Nishino, M.N., Tsunakawa, H., Shibuya, H., Matsushima, M., Shimizu, H., Takahashi, F., Fujimoto, M., Mukai, T., Terasawa, T.: First in situ observation of the Moon-originating ions in the Earth's Magnetosphere by MAP-PACE on SELENE (KAGUYA). Geophys. Res. Lett. **36**, L22106 (2009). doi:10.1029/2009GL040682
169. Tao, J.B., Ergun, R.E., Newman, D.L., Halekas, J.S., Andersson, L., Angelopoulos, V., Bonnell, J.W., McFadden, J.P., Cully, C.M., Auster, H.U., Glassmeier, K.H., Larson, D.E., Baumjohann, W., Goldman, M.V.: Kinetic instabilities in the lunar wake: ARTEMIS observations. J. Geophys. Res. **117**, A03106 (2012). doi:10.1029/2011JA017364
170. Thompson, M.W.: The energy spectrum of ejected atoms during the high energy sputtering of gold. Philos. Mag. 18(152), 377–414 (1968). doi:10.1080/14786436808227358. http://www.tandfonline.com/doi/abs/10.1080/14786436808227358
171. Townsend, P.C.: Sputtering by electrons and photons. Sputtering Particle Bombardment II **52**, 147–178 (1983)
172. Tsugawa, Y., Katoh, Y., Terada, N., Ono, T., Tsunakawa, H., Takahashi, F., Shibuya, H., Shimizu, H., Matsushima, M., Saito, Y., Yokota, S., Nishino, M.N.: Statistical study of broadband whistler-mode waves detected by Kaguya near the Moon. Geophys. Res. Lett. **39**, L16101 (2012). doi:10.1029/2012GL052818
173. Tsugawa, Y., Terada, N., Katoh, Y., Ono, T., Tsunakawa, H., Takahashi, F., Shibuya, H., Shimizu, H., Matsushima, M.: Statistical analysis of monochromatic whistler waves near the Moon detected by Kaguya. Ann. Geophys. **29**(5), 889–893 (2011). doi:10.5194/angeo-29-889-2011
174. Tsurutani, B.T.: Comets: a laboratory for plasma waves and instabilities. Geophys. Monogr. Ser. **61**, 189–209 (1991). doi:10.1029/GM061p0189
175. Umeda, T.: Effect of ion cyclotron motion on the structure of wakes: a vlasov simulation. Earth Planets Space **64**, 231–236 (2012)
176. Van Allen, J.A., Ness, N.F.: Particle shadowing by the Moon. J. Geophys. Res. **74**, 71–93 (1969). doi:10.1029/JA074i001p00071
177. Vorburger, A., Wurz, P., Barabash, S., Wieser, M., Futaana, Y., Holmström, M., Bhardwaj, A., Asamura, K.: Energetic neutral atom observations of magnetic anomalies on the lunar surface. J. Geophys. Res. **117**, A07208 (2012). doi:10.1029/2012JA017553
178. Vorburger, A., Wurz, P., Barabash, S., Wieser, M., Futaana, Y., Lue, C., Holmström, M., Bhardwaj, A., Dhanya, M.B., Asamura, K.: Energetic neutral atom imaging of the lunar surface. J. Geophys. Res. **118**, 3937–3945 (2013). doi:10.1002/jgra.50337. http://dx.doi.org/10.1002/jgra.50337
179. Wang, X., Horányi, M., Robertson, S.: Characteristics of a plasma sheath in a magnetic dipole field: Implications to the solar wind interaction with the lunar magnetic anomalies. J. Geophys. Res. **117**, A06226 (2012). doi:10.1029/2012JA017635
180. Wang, X.D., Zong, Q.G., Wang, J.S., Cui, J., Rème, H., Dandouras, I., Aoustin, C., Tan, X., Shen, J., Ren, X., Liu, J.J., Zuo, W., Su, Y., Wen, W.B., Wang, F., Fu, Q., Mu, L.L., Wang, X.Q., Geng, L., Zhang, Z.B., Liu, J.Z., Zhang, H.B., Li, C.L., Ouyang, Z.Y.: Detection of m/q = 2 pickup ions in the plasma environment of the Moon: The trace of exospheric $H_2{}^+$. Geophys. Res. Lett. **38**, L14204 (2011). doi:10.1029/2011GL047488

181. Whang, Y.C.: Interaction of the magnetized solar wind with the moon. Phys. Fluids **11**, 969–975 (1968). doi:10.1063/1.1692068
182. Whang, Y.C.: Theoretical study of the magnetic field in the lunar wake. Phys. Fluids **11**, 1713–1719 (1968). doi:10.1063/1.1692185
183. Whipple, E.C.: Potentials of Surfaces in Space. Rep. Prog. Phys. **44**(11), 1197–1250 (1981). doi:10.1088/0034-4885/44/11/002. http://stacks.iop.org/0034-4885/44/i=11/a=002
184. Wiehle, S., Plaschke, F., Motschmann, U., Glassmeier, K.H., Auster, H.U., Angelopoulos, V., Mueller, J., Kriegel, H., Georgescu, E., Halekas, J., Sibeck, D.G., McFadden, J.P.: First lunar wake passage of ARTEMIS: Discrimination of wake effects and solar wind fluctuations by 3D hybrid simulations. Planet. Space Sci. **59**(8), 661–671 (2011). doi:10.1016/j.pss.2011.01.012
185. Wiens, R.C., Burnett, D.S., Calaway, W.F., Pellin, M.J.: Experimental studies of the role of photodesorption in the formation of planetary Na atmospheres. Bull. Am. Astron. Soc. **25**, 1089 (1993)
186. Wieser, M., Barabash, S., Futaana, Y., Holmström, M., Bhardwaj, A., Sridharan, R., Dhanya, M.B., Schaufelberger, A., Wurz, P., Asamura, K.: First observation of a mini-magnetosphere above a lunar magnetic anomaly using energetic neutral atoms. Geophys. Res. Lett. **37**, L05103 (2010). doi:10.1029/2009GL041721
187. Wieser, M., Barabash, S., Futaana, Y., Holmström, M., Bhardwaj, A., Sridharan, R., Dhanya, M.B., Wurz, P., Schaufelberger, A., Asamura, K.: Extremely high reflection of solar wind protons as neutral hydrogen atoms from regolith in space. Planet. Space Sci. **57**(14–15), 2132–2134 (2009). doi:10.1016/j.pss.2009.09.012
188. Winslow, R.M., Johnson, C.L., Anderson, B.J., Korth, H., Slavin, J.A., Purucker, M.E., Solomon, S.C.: Observations of Mercury's northern cusp region with MESSENGER's Magnetometer. Geophys. Res. Lett. **39**, L08112 (2012). doi:10.1029/2012GL051472
189. Wurz, P., Lammer, H.: Monte-carlo simulation of mercury's exosphere. Icarus **164**(1), 1–13 (2003). doi:10.1016/S0019-1035(03)00123-4. http://www.sciencedirect.com/science/article/pii/S0019103503001234
190. Yokota, S., Saito, Y.: Estimation of picked-up lunar ions for future compositional remote SIMS analyses of the lunar surface. Earth Planets Space **57**(4), 281–289 (2005)
191. Yokota, S., Saito, Y., Asamura, K., Tanaka, T., Nishino, M.N., Tsunakawa, H., Shibuya, H., Matsushima, M., Shimizu, H., Takahashi, F., Fujimoto, M., Mukai, T., Terasawa, T.: First direct detection of ions originating from the Moon by MAP-PACE IMA onboard SELENE (KAGUYA). Geophys. Res. Lett. **36**, L11201 (2009). doi:10.1029/2009GL038185
192. Zhang, H., Khurana, K.K., Zong, Q.G., Kivelson, M.G., Hsu, T.S., Wan, W.X., Pu, Z.Y., Angelopoulos, V., Cao, X., Wang, Y.F., Shi, Q.Q., Liu, W.L., Tian, A.M., Tang, C.L.: Outward expansion of the lunar wake: Artemis observations, Geophys. Res. Lett. **39**, L18104 (2012). doi:10.1029/2012GL052839
193. Zhou, X.Z., Angelopoulos, V., Poppe, A.R., Halekas, J.S.: Artemis observations of lunar pick-up ions: mass constraints on ion species. J. Geophys. Res. **118**, 1766–1774 (2013). doi:10.1002/jgre.20125. http://dx.doi.org/10.1002/jgre.20125
194. Zimmerman, M.I., Farrell, W.M., Stubbs, T.J., Halekas, J.S., Jackson, T.L.: Solar wind access to lunar polar craters: feedback between surface charging and plasma expansion. Geophys. Res. Lett. **38**, L19202 (2011). doi:10.1029/2011GL048880
195. Zimmerman, M.I., Jackson, T.L., Farrell, W.M., Stubbs, T.J.: Plasma wake simulations and object charging in a shadowed lunar crater during a solar storm. J. Geophys. Res. **117**, E00K03 (2012). doi:10.1029/2012JE004094
196. Zurbuchen, T.H., Raines, J.M., Slavin, J.A., Gershman, D.J., Gilbert, J.A., Gloeckler, G., Anderson, B.J., Baker, D.N., Korth, H., Krimigis, S.M., Sarantos, M., Schriver, D., McNutt Jr, R.L., Solomon, S.C.: MESSENGER observations of the spatial distribution of planetary ions near Mercury. Science **333**(6051), 1862–1865 (2011). doi:10.1126/science.1211302

Chapter 2
Electron Gyro-Scale Dynamics Near the Lunar Surface

Abstract I examine electron dynamics near the lunar surface with spatial scale lengths characterized by the electron gyroradius from both observational and theoretical standpoints. Electron velocity distribution functions (VDFs) obtained by Kaguya at ∼10–100 km altitudes above the Moon occasionally exhibit clear non-gyrotropic signatures, which have been rarely observed in space plasma. Comparison of the observed electron VDFs with theoretical predictions derived from particle-trace calculations demonstrates that electron absorption by the lunar surface combined with electron gyromotion produces a gyrophase-dependent (non-gyrotropic) partial loss in the electron VDF. The electron motion is modified by the near-lunar electromagnetic environment, and therefore, the characteristics of the non-gyrotropic electron VDFs reflect the electric and magnetic fields in the vicinity of the Moon. I discuss the Moon-related electromagnetic fields by analyzing the high-angular resolution electron data from Kaguya.

Keywords Non-gyrotropic electrons · Electron gyromotion · Lunar magnetic anomaly · Non-adiabatic scattering

2.1 Introduction

This chapter discusses electron gyro-scale dynamics in the vicinity of the Moon from Kaguya observations and test-particle tracing, including two main topics: (i) non-gyrotropic electron velocity distribution functions, and (ii) non-adiabatic magnetic scattering of electrons. First I introduce basic knowledge for each topic.

2.1.1 Non-gyrotropic Velocity Distribution Functions

In thermal equilibrium, ions and electrons are expected to have gyrotropic (i.e., symmetric relative to the magnetic-field lines) velocity distribution functions (VDFs) in the plasma's rest frame. Plasma in space is not necessarily in thermal equilibrium, and ions and electrons occasionally form non-gyrotropic VDFs associated with

Y. Harada, *Interactions of Earth's Magnetotail Plasma with the Surface, Plasma, and Magnetic Anomalies of the Moon*, Springer Theses,
DOI 10.1007/978-4-431-55084-6_2

shocks, boundaries, or waves. Three possible mechanisms to generate non-gyrotropic (gyrophase-bunched) ions include (i) a pick-up process of ions by the solar wind, (ii) plasma inhomogeneities on spatial scales smaller than the ion gyroradius, and (iii) wave–particle interactions (resonant trapping by waves). Non-gyrotropic ions possibly produced by these mechanisms have been widely observed near the Earth's bow shock and interplanetary shocks, in the current sheet and the plasma-sheet boundary layer in the Earth's magnetotail, and near comets, planets, the Moon, and in similar environments [3, 4, 11, 26, 28, 40].

In contrast to the large number of observations of non-gyrotropic *ion* VDFs, only a few observational papers exist that focus on non-gyrotropic *electron* VDFs. The small electron gyroradius and short gyroperiod compared with the ion gyroradius and gyroperiod cause great difficultly in attempts to detect non-gyrotropic electron VDFs [19]. Some signatures of a gyrophase-bunched electron VDF were observed just upstream of the Earth's bow shock, first by ISEE 1 and ISEE 2, and subsequently by WIND [1, 12]. Although a local phase-trapping distribution is thought to be necessary for gyrophase-bunched electron observations, the mechanism responsible for producing these non-gyrotropic electrons is as yet unknown.

2.1.2 Lunar Magnetic Field Measurements

The Moon does not possess a global intrinsic magnetic field, but it has remanent crustal magnetic fields distributed widely and non-uniformly over the lunar surface [8] that are known as magnetic anomalies. Lunar crustal magnetization has been extensively studied for decades but its origins remain one of the key unsolved problems in the study of lunar history [6]. Critical hurdles include the difficulty in mapping lunar magnetic fields with a wide range of spatial scales from a few kilometers or less up to hundreds of kilometers.

Direct measurements of lunar crustal magnetic fields are provided by magnetometers installed on lunar orbiting spacecraft as well as by those deployed on the lunar surface. When a lunar orbiter travels over a magnetic anomaly, the magnetometer detects Moon-related field variations, which are fixed to the selenographic coordinate system. After detrending the external field variations, I can map the global distribution of the three components of the lunar crustal magnetic fields [5, 16, 18, 29–31, 38, 39]. In practice, the orbital altitude, usually tens of kilometers or more, limits the detectable spatial scale length of the crustal fields. The strength of small-scale magnetic fields distributed on the surface decreases sharply with altitude and eventually becomes smaller than the magnetometer noise level at the orbital altitude. Lunar surface magnetometers made direct measurements of the surface magnetic fields at the Apollo and Lunokhod-2 landing sites [7, 8]. Strong crustal magnetic fields up to $\sim$327 nT were measured, and both the field strength and polarity vary significantly over a few kilometers or less. The surface magnetic field data are only available for a few landing sites, and it remains uncertain whether kilometer-scale magnetization exists elsewhere on the lunar surface.

Another approach is indirect estimation of crustal magnetic fields using charged-particle motion in the vicinity of magnetized regions. The most frequently used remote sensing technique is "electron reflectometry" (ER), which is based on the magnetic mirror effect [2, 15, 20, 21, 27]. The ER technique assumes adiabatic behavior, which means that the field variation encountered by electrons within a single gyration orbit is small compared with the initial field. This assumption is not appropriate for magnetic fields with spatial scales smaller than the electron gyrodiameter (200 eV electrons have a gyrodiameter of 9.5 km in a 10 nT magnetic field), and crustal-field strengths could be significantly underestimated by the ER technique in the presence of kilometer-scale magnetization [14]. Recently, magnetic anomaly maps were produced from the observations of solar wind protons deflected by strong crustal magnetic fields [23] and from those of energetic neutral atoms formed from solar wind protons, which are backscattered as neutral hydrogen atoms at the surface [41]. Since these two techniques are based on the ion dynamics, they are appropriate for mapping surface magnetic fields with length scales comparable to or greater than the ion gyrodiameters.

2.2 Instrumentation and Coordinates

Kaguya is a lunar orbiter that was launched on 14 September 2007 and entered a circular lunar polar orbit with an altitude of 100 km. Since Kaguya is a three-axis stabilized spacecraft, one of its panels always faces the lunar surface. Its altitude was lowered to ~50 km from January 2009 to April 2009, and the lowest periapsis altitude was ~10 km after April 2009 until it hit the lunar surface on 10 June 2009. In this chapter, I analyze data obtained by the MAgnetic field and Plasma experiment (MAP) instrument on Kaguya, which consists of two components: the Lunar MAGnetometer (LMAG) and the Plasma energy Angle and Composition Experiment (PACE). LMAG is a triaxial fluxgate magnetometer, which is used to observe the magnetic field with a sampling frequency of 32 Hz and a resolution of 0.1 nT [24, 36, 37, 39]. Here 1 s magnetic-field data are used. Low-energy charged particles near the Moon are observed by PACE, which consists of four sensors: two electron spectrum analyzers (ESA-S1 and ESA-S2), an ion mass analyzer (IMA), and an ion energy analyzer (IEA) [34, 35]. ESA-S1 and ESA-S2 measure the distribution functions of low-energy electrons in the energy ranges of 6 eV–9 keV and 9 eV–16 keV, respectively. Each electron sensor has a hemispherical field of view, with angular resolutions of a 5° full width at half maximum (FWHM) in elevation and an 8° FWHM in azimuthal angle. I use the satellite coordinates; i.e., $+Z$ is directed towards the lunar surface, $+X$ or $-X$ is the direction of travel, and Y completes the orthogonal coordinate set (see Fig. 2.1). ESA-S1 and IMA are installed on the $+Z$ panel (looking down toward the lunar surface), while ESA-S2 and IEA are on the $-Z$ panel (looking away from the lunar surface). I define the azimuthal angle of the magnetic field in the satellite coordinates as ϕ_{Bsat} and the corresponding elevation angle as θ_{Bsat}.

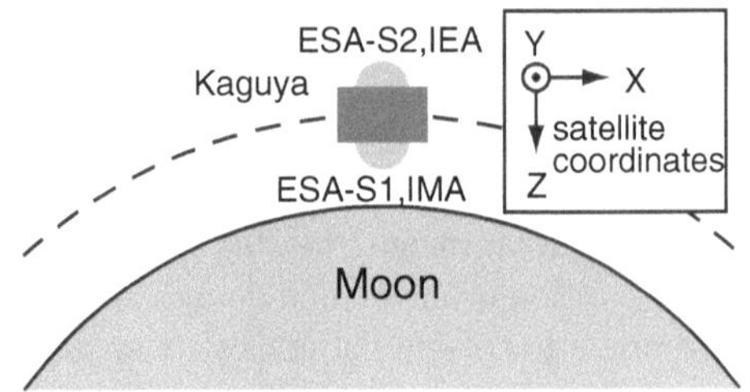

Fig. 2.1 Satellite coordinates of Kaguya. $+Z$ is directed toward the lunar surface; $+X$ or $-X$ is the direction of travel; and Y completes the orthogonal set

2.3 Non-gyrotropic Electron Velocity Distribution Functions Near the Lunar Surface

This section presents Kaguya observations of non-gyrotropic electron VDFs near the lunar surface as well as theoretical considerations of their generation mechanisms. These near-lunar non-gyrotropic electrons are produced by the "gyro-loss" effect; surface absorption of electrons combined with their gyromotion gives rise to a non-gyrotropic partial loss in the electron VDF observed by spacecraft with altitudes lower than the electron gyrodiameter. The gyro-loss effect is essentially the same as "cyclotron shadowing" considered in the Apollo era. Shadowing of gyrating particles by an absorbing surface was first theoretically treated by calculating the reduction in omnidirectional particle flux observed by lunar orbiters [25]. This concept was then applied to ion and electron measurements by a detector deployed on the lunar surface with a given look direction [32]. In addition, the local crustal magnetic field at the Apollo 14 site was taken into account in the numerical calculations, and it was shown that the theoretical calculations were consistent with a "magnetic shadowing" observation of plasma sheet electrons by Charged Particle Lunar Environment Experiment (CPLEE), which consists of two small-aperture detectors [33]. Kaguya's high-angular-resolution measurement of 3D electron VDF provides us clear observational evidence of cyclotron shadowing in low lunar orbits.

I first consider an ideal and simple case of a uniform magnetic field parallel to a plane surface to understand the basic nature of the electron dynamics at low altitudes above the lunar surface. Test particle tracing is then conducted to take into account more realistic and complicated configurations, including the spherical lunar surface, inclined and nonuniform magnetic fields, the negatively charged lunar surface, perpendicular electric fields, etc. I compare the theoretical predictions with the electron VDFs obtained by Kaguya, discussing how the electron motion is modified by the lunar electromagnetic environment. In this thesis, I refer to "empty regions" when I am describing features in observations, while "forbidden regions" when I am describing theoretical predictions. Note that here I refer to the "forbidden regions" in velocity phase space, wheares this term was originally used by Størmer to describe the spatial regions of a dipole magnetic field to which charged particles from the sun do not have direct access [17].

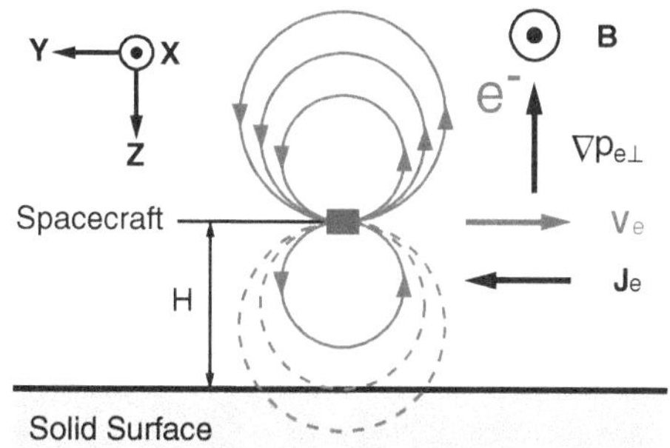

Fig. 2.2 Schematic illustration of the gyro-loss effect and the diamagnetic current

2.3.1 Gyro-Loss Effect on Electron Velocity Distribution Functions

2.3.1.1 Theoretical Predictions

A charged particle gyrates around magnetic-field lines with a gyroradius (Larmor radius) given by $r_L = mv_\perp/|q|B$, where m is the particle mass, $v_\perp$ the velocity component perpendicular to the magnetic field, q the electric charge, and B the magnetic-field intensity. Consider a VDF at an altitude H in the presence of a magnetic field oriented parallel to a plane solid surface (cf. Fig. 2.2). As shown by the gray dashed lines in Fig. 2.2, some particles on orbits with gyrodiameters greater than or equal to H strike the surface and are absorbed because of gyromotion. This absorption results in an empty region in the VDF. This gyro-loss effect is expected to be seen in electron VDFs observed by Kaguya, because the gyrodiameter of electrons near the Moon (1 keV electrons have a gyrodiameter of 107 km in a 2 nT magnetic field) is comparable in length to Kaguya's orbital height (~10–100 km).

When an electron enters the sensors with a perpendicular velocity component $v_\perp$ and a gyrophase ψ as illustrated in Fig. 2.3, a critical gyroradius is given by

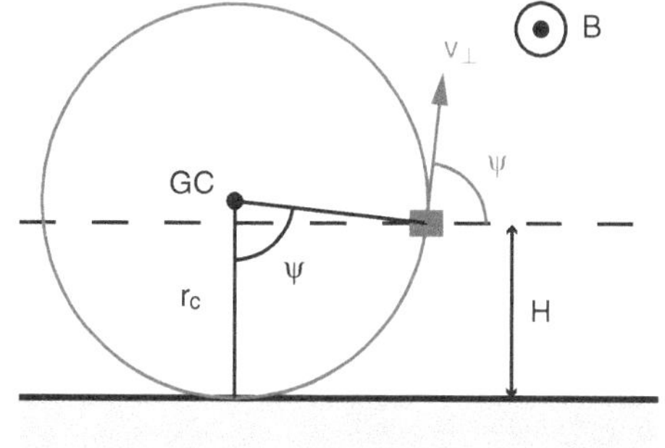

Fig. 2.3 Schematic illustration of the critical gyroradius, r_c, gyrophase, ψ, perpendicular velocity, $v_\perp$, spacecraft orbital height, H, and the guiding center of the electron, GC

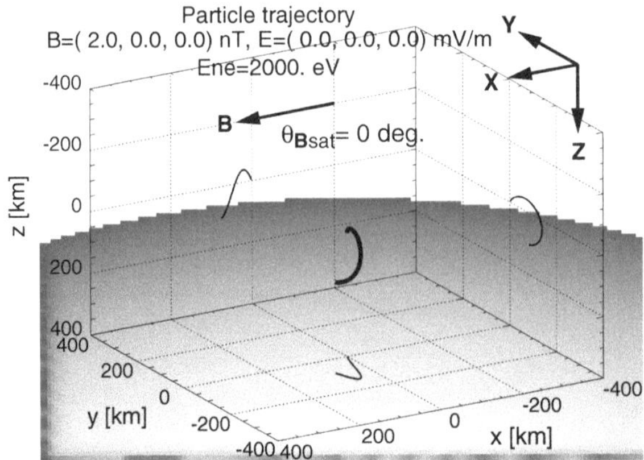

Fig. 2.4 Sample electron trajectories obtained from reverse tracking calculations, launched from a spacecraft altitude of 100 km with an energy of 2 keV, for a uniform magnetic field of 2 nT and no electric field; the *black trace* shows the trajectory for $\theta_{\rm Bsat} = 0°$

$$r_{\rm c} = \frac{H}{1 - \cos\psi}. \tag{2.1}$$

Electrons with $r_{\rm L} \geq r_{\rm c}$ strike the lunar surface and therefore cannot be observed. When $\psi = 180°$, r_c takes a minimum value $H/2$ and the cut-off energy of electrons will be a minimum. On the other hand, r_c is infinite when $\psi = 0°$ and no electrons will be cut off for this angle. Therefore, whether or not an electron is absorbed by the lunar surface depends on its gyrophase, resulting in a gyrophase-bunched (non-gyrotropic) partial loss in the electron VDF.

To consider the gyro-loss effect on electron VDFs observed by Kaguya, I perform particle-trace calculations using a fourth-order Runge–Kutta integration method. I backtrace a test electron from the spacecraft and check whether or not it strikes the lunar surface, assuming the Moon to be a sphere with a radius of $1\,R_{\rm L}$ ($1\,R_{\rm L} =$ 1738 km). The black trace in Fig. 2.4 shows an example of an electron trajectory that results in a collision with the lunar surface in the presence of a uniform magnetic field that is locally parallel to the lunar surface at the spacecraft's position ($\theta_{\rm Bsat} = 0°$) and in the absence of electric fields. I launch electrons characterized by an initial velocity distribution divided into 64 elevation angles, 64 azimuthal angles, and four different kinetic energies. This way, I derive the forbidden regions in the electron VDFs. Electrons launched with velocities in these forbidden regions strike the lunar surface.

For a uniform magnetic field that is locally parallel to the lunar surface at the spacecraft's locus and in the absence of electric fields, the derived forbidden regions are shown in white in Fig. 2.5. I trace electrons with a time step of 1/50 of the gyroperiod T_{ce} from the spacecraft's altitude of 100 km until they either strike the lunar surface or gyrate one cycle, $t \geq -T_{ce}$. (One cycle is sufficient to see whether an electron strikes the Moon in the presence of a parallel magnetic field.) In Fig. 2.5, the

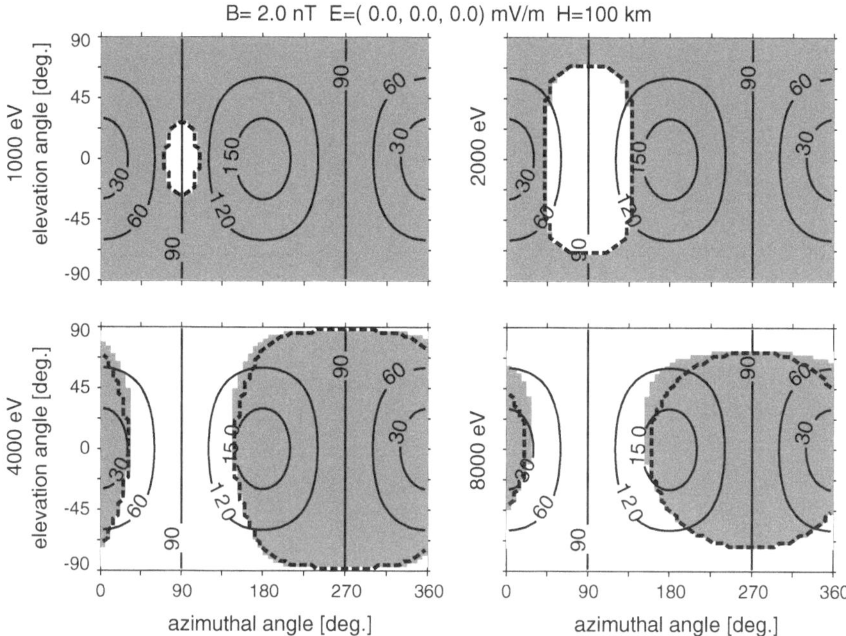

Fig. 2.5 Forbidden regions in the electron VDF for different energies in satellite coordinates, assuming a uniform magnetic field of 2 nT in the $+X$ direction, no electric fields, and a spacecraft altitude of 100 km. The *contours* indicate pitch angles. The *white regions* show the forbidden regions derived from particle-trace calculations. The *dashed lines* indicate the forbidden regions derived from Eq. (2.1)

high-energy forbidden regions are larger than their low-energy counterparts, since more electrons strike the lunar surface because of the larger gyroradii associated with higher energies. The forbidden regions appear at an azimuthal angle of around 90° and an elevation angle of approximately 0°. This angle corresponds to $\psi = 180°$, where electrons strike the lunar surface with the minimum energy.

The dashed lines in Fig. 2.5 indicate the forbidden regions derived from Eq. (2.1). These regions are symmetric with respect to the (v_x, v_y) plane, since the lunar surface is assumed to be planar. By contrast, the forbidden regions derived from the particle-trace calculation (white regions in Fig. 2.5) are asymmetric relative to the (v_x, v_y) plane since I assume the Moon to be a sphere: the time in which electrons that are launched for reverse tracking with initial z-component velocity $v_z > 0$ (the regions of positive elevation angle in Fig. 2.5 and $180 < \psi < 360°$ in Fig. 2.3) gyrate from the spacecraft to the lunar surface is longer than for electrons with $v_z < 0$. Therefore, electrons with $v_z > 0$ can move long distances along the magnetic-field lines (thus increasing their effective height because of the spherical Moon) and the forbidden regions for $v_z > 0$ are smaller than those for $v_z < 0$.

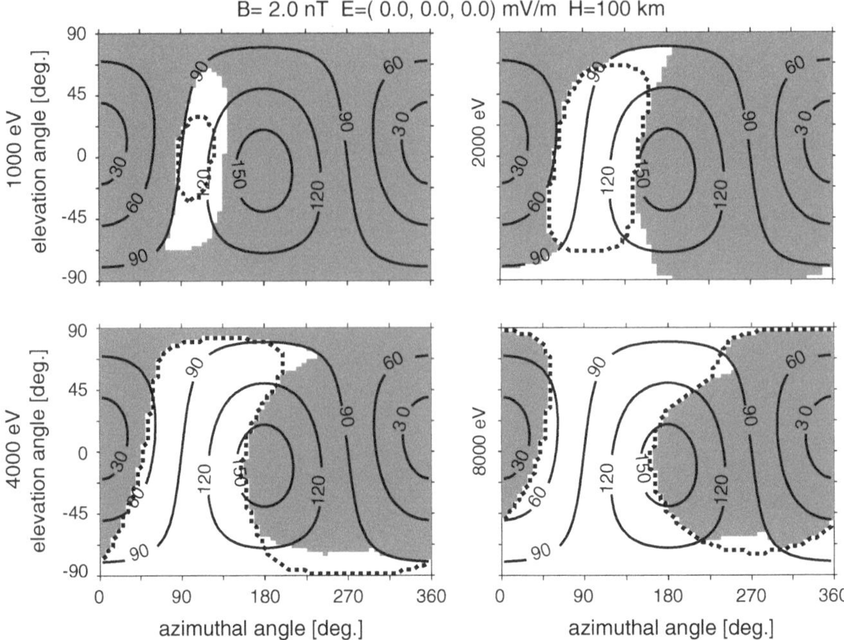

Fig. 2.6 Forbidden regions in the electron VDF for an inclined magnetic field ($\theta_{\text{Bsat}} = +10°$), in the same format as in Fig. 2.5. The *dashed lines* indicate the regions where electrons strike the lunar surface during one gyrocycle

For magnetic-field lines that pass through the spacecraft under an angle but do not intersect the lunar surface, the derived forbidden regions are shown in Fig. 2.6. I perform this calculation by adopting $\theta_{\text{Bsat}} = +10°$ (the magnetic-field line at 100 km intersects the Moon with $\theta_{\text{Bsat}} \geq +19°$). I continuously trace electrons until they either strike the lunar surface or gyrate twenty cycles: $t \geq -20T_{ce}$. The dashed lines in Fig. 2.6 indicate the regions where the electrons strike the surface during one cycle. Although the forbidden regions in Fig. 2.6 also appear at an azimuthal angle of around 90° and an elevation angle of approximately 0°, they have different distributions from those shown in Fig. 2.5. The forbidden regions are large for pitch angles $\alpha > 90°$, since electrons gyrate upwards along the magnetic-field lines. In the forbidden regions with $\alpha > 90°$, electrons launched from the spacecraft for reverse tracking can strike the lunar surface after they gyrate more than one cycle, as indicated by the black trace in Fig. 2.7. On the other hand, electrons in the forbidden regions with $\alpha < 90°$ strike the lunar surface in one gyrocycle, because they move away from the lunar surface during reverse tracking. Electrons with $\alpha > 90°$ pass near the lunar surface and can strike with lower energies than electrons with $\alpha < 90°$ if $\theta_{\text{Bsat}} > 0$.

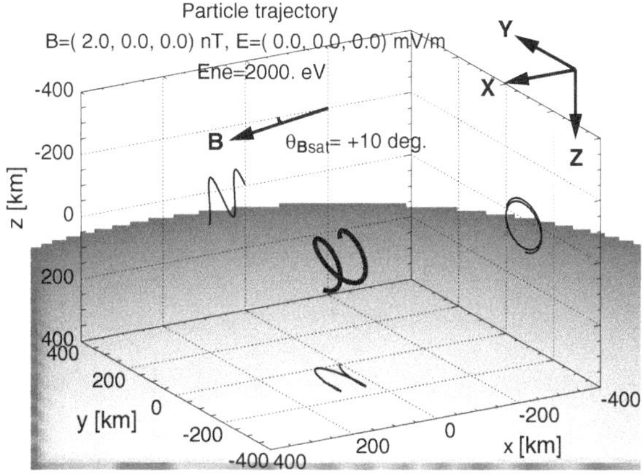

Fig. 2.7 Sample electron trajectories obtained from reverse tracking calculations, launched from a spacecraft altitude of 100 km with an energy of 2 keV, for a uniform magnetic field of 2 nT and no electric field; the *black trace* shows the trajectory for $\theta_{\mathrm{Bsat}} = +10°$

2.3.1.2 Observations

Here I present four electron VDFs which exhibit non-gyrotropic empty regions produced by the gyro-loss effect. Table 2.1 shows the location of the Moon and Kaguya, as well as the ambient plasma conditions for these events.

Table 2.1 Plasma conditions for four events

	Event 1	Event 2	Event 3	Event 4
Time period	2009/06/05	2009/04/17	2008/04/18	2008/01/21
	20:10:17–20:10:33	11:57:22–11:58:42	23:06:33–23:07:21	15:00:23–15:01:11
Moon location at GSE coordinates	$(-58, 23, -5)\ R_{\mathrm{E}}$	$(-1, -63, -1) R_{\mathrm{E}}$	$(-60, 18, -4) R_{\mathrm{E}}$	$(-57, 12, 3) R_{\mathrm{E}}$
Kaguya location				
• Orbital height (km)	12	14	109	98
• Latitude and longitude in selenographic coordinates	(76°S, 149°E)	(36°S, 104°W)	(62°N, 176°W)	(28°S, 92°W)
• Solar zenith angle	99°	39°	115°	106°
Magnetic field intensity (nT)	4.6	5.1	2.1	2.1
$(\phi_{\mathrm{Bsat}}, \theta_{\mathrm{Bsat}})$	(41°, 2°)	(268°, −2°)	(338°, −2°)	(239°, −11°)
Density (cm^{-3})	0.10	3.20	0.06	0.07
Ion temperature (IEA)	–	27 eV	1.93 keV	1.86 keV
Electron temperature (eV)	88	21	446	435
Bulk flow (IEA) (km/s)	–	446	490	410

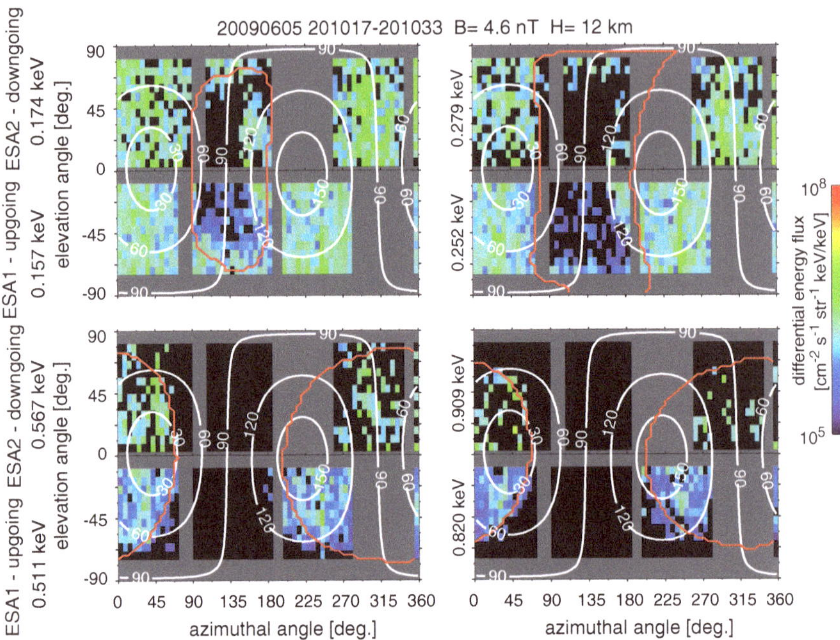

Fig. 2.8 Electron angular distribution for different energies in satellite coordinates obtained in the period of 20:10:17–20:10:33 UT (16 s) on 5 June 2009, when the Moon was in the Earth's magnetosphere. Angles with little or no sensitivity are indicated in *gray*. Note that ESA-S1 and ESA-S2 have different sensitivities. The *white contours* represent the pitch angles. The *red solid lines* indicate the forbidden regions derived from particle-trace calculations for the energies of ESA-S1 and ESA-S2 shown in each panel, assuming a uniform magnetic field and no electric field

Figure 2.8 shows a gyro-loss event, observed at an altitude of only 12 km (time-series data are shown in Fig. 2.9). At this time, the Moon was thought to be located in either the plasma-sheet boundary layer or the magnetotail lobe, and Kaguya was located on the nightside of the Moon near the south pole (at latitude 76° S, longitude 149° E in selenographic coordinates), where relatively weak and/or small-scale magnetic anomalies exist [39]. The red solid lines in Fig. 2.8 show the forbidden regions derived from the particle-trace calculation using the magnetometer data and assuming an inclined, uniform magnetic field without any electric fields. Empty regions clearly appeared at an azimuthal angle of around 135°. They roughly correspond to the theoretically derived forbidden regions indicated by the red solid lines, although the empty regions seem to be somewhat smaller than the forbidden regions. At lower altitudes of ~10 km, the gyro-loss effect is clearly seen, even for lower energies of a few hundred eV.

Figure 2.10 shows an electron VDF observed during Event 2, when the Moon was located in the solar wind and Kaguya's position was on the far side (at latitude 36° S, longitude 104° W in selenographic coordinates), where strong magnetic anomalies do not exist [39]. Detailed information and time-series data for Event 2 are included

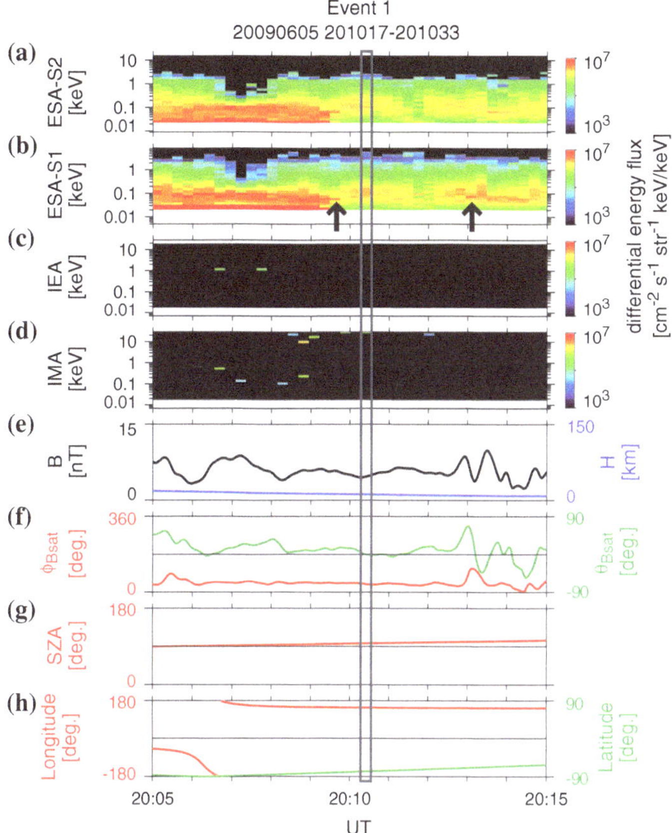

Fig. 2.9 Time-series data from Kaguya for Event 1. **a–d** Energy–time spectrograms from PACE sensors, **e** magnetic-field intensity and Kaguya's orbital altitude, **f** direction of magnetic field in satellite coordinates, and **g** and **h** spacecraft location in solar zenith angle (SZA) and in selenographic coordinates. The *gray box* indicates the time period of the gyro-loss event shown in Fig. 2.8. The *black arrows* in panel **b** indicate upward-going electron beams with energies of ~50 eV, observed by ESA-S1

in Table 2.1 and shown in Fig. 2.11. In the solar wind, the electron temperature is significantly lower than that in the terrestrial plasma sheet, and observed counts of higher-energy electrons are very small. Therefore, I did not observe the gyro-loss effect at higher altitudes and even at lower altitudes, I have to average counts over a certain interval to see the gyro-loss effect on electron VDFs in the solar wind. Figure 2.10 shows the electron distribution averaged during the period covering 11:57:22–11:58:42 UT on 17 April 2009, when the magnetic field was quite stable and nearly parallel to the lunar surface. The electron flux clearly decreases in the theoretically derived forbidden regions indicated by the red lines in Fig. 2.10. This shows the gyro-loss effect on electron VDFs also when the Moon is located in the

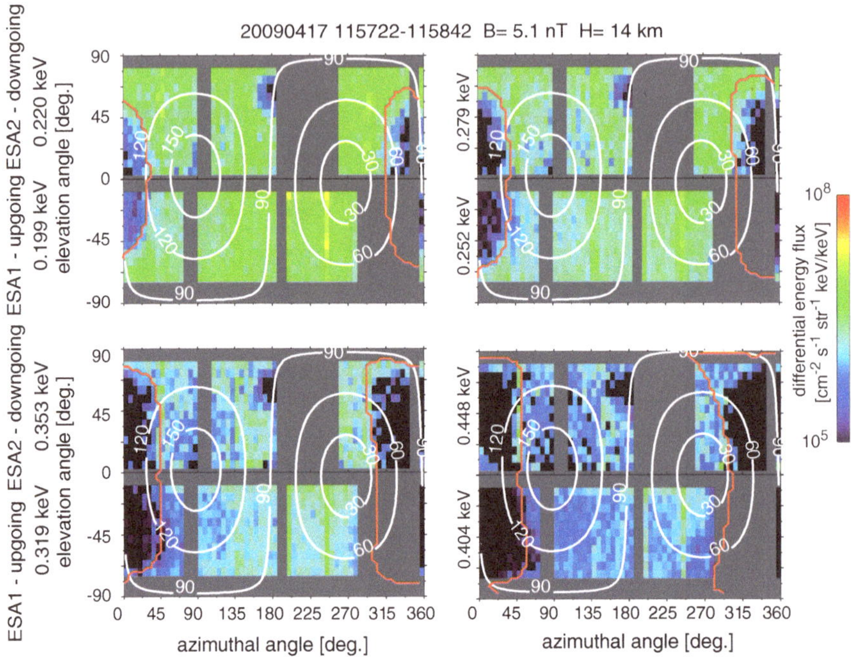

Fig. 2.10 Electron angular distribution obtained during 11:57:22–11:58:42 UT (80 s average) on 17 April 2009, when the Moon was located in the solar wind, in the same format as in Fig. 2.8. The *red lines* indicate the forbidden regions for a uniform magnetic field and no electric field. The flux dips around (180°, 65°) are thought to be due to blockage of electrons by the high-gain antenna

solar wind, although there are slight differences between these red lines and the boundaries of the observed empty regions.

I have shown the two gyro-loss events observed in both the Earth's magnetosphere and the solar wind. The observed electron VDFs are relatively consistent with theoretical predictions for uniform magnetic fields without electric fields. However, I also found some distributions exhibiting empty regions that do not correspond to the theoretically derived forbidden regions based on this simple assumption. Two examples that have remarkably different distributions from the simple theoretical predictions are shown in Figs. 2.12 and 2.13 (detailed information and time-series data are included in Table 2.1 and shown in Figs. 2.14 and 2.15). Both of these events were observed when Kaguya was located above the rather weak and/or small-scale crustal-field regions [39], and the Moon was located in the high-temperature and weak-magnetic-field regions in the central plasma sheet. The electron VDFs of these events indicate that large numbers of electrons were observed in the forbidden regions indicated by the red lines. The electron motions and forbidden regions may have been greatly modified by some factors, including electric and nonuniform magnetic fields. I do not deal with the effects of lunar magnetic anomalies in this section, although they may play an important role, in particular around strong magnetic anomalies.

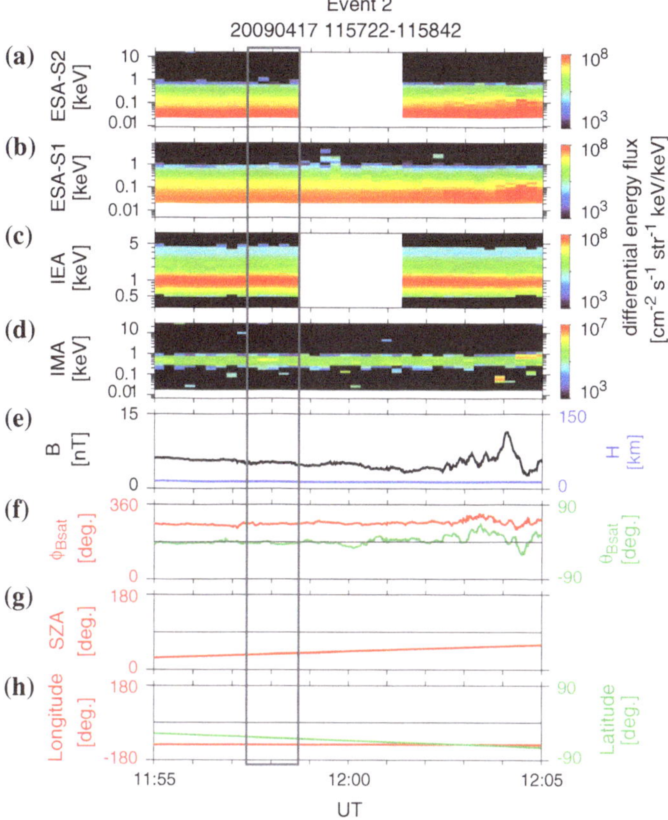

Fig. 2.11 Time-series data from Kaguya for Event 2 in the same format as in Fig. 2.9. The *gray box* indicates the time period of the gyro-loss event shown in Fig. 2.10

They will be discussed in the next section. Nonuniform magnetic fields caused by a diamagnetic-current system near the lunar surface, lunar-surface charging, and electric fields perpendicular to the magnetic field are discussed below.

2.3.2 Diamagnetic-Current System

Non-gyrotropic electron VDFs produced by the gyro-loss effect are related to a diamagnetic-current system formed by the pressure gradient near the lunar surface. The diamagnetic current produces magnetic fields, resulting in nonuniform magnetic-field structures. The nonuniform magnetic fields will modify the electron trajectories and, consequently, the forbidden regions. This diamagnetic-current effect will contribute strongly in the presence of high plasma pressure and weak magnetic fields.

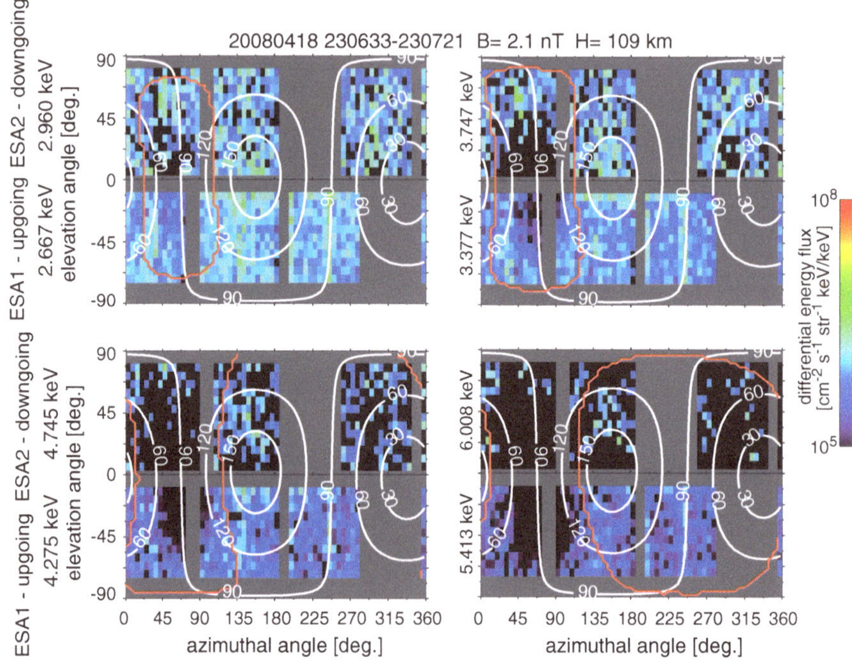

Fig. 2.12 Electron angular distribution obtained during 23:06:33–23:07:21 UT (48 s average) on 18 April 2008, when the Moon was in the Earth's magnetosphere, in the same format as in Fig. 2.8. The *red lines* indicate the forbidden regions for a uniform magnetic field and no electric field

The following is discussion of how the diamagnetic-current system affects the forbidden regions in electron VDFs.

When I consider the lunar surface and a magnetic field that is parallel to the surface, high-energy, hot electrons are gradually lost as the altitude H decreases because of the gyro-loss effect, and the pressure gradient near the surface forms a diamagnetic-current system (see Fig. 2.2). Although numerous hot electrons are lost near the surface, charge neutrality would be preserved by field-aligned intrusion of cold electrons, which are not lost through the gyro-loss effect because of their small gyroradii.

For the ambient plasma near the Moon, most ions have gyroradii that are much larger than H or even larger than the lunar radius, except when the Moon is in the magnetotail lobes. These ions are not magnetized on the scale length considered here, and I do not take into account the ion diamagnetic current near the lunar surface.

In the presence of a gradient of perpendicular pressure to the magnetic field, $\nabla p_{e\perp}$, the diamagnetic electron current is given by

$$\mathbf{J}_{\mathrm{D}e} = \frac{\mathbf{B} \times \nabla p_{e\perp}}{B^2} = \frac{\mathbf{B} \times \nabla (n_{e\mathrm{h}} k_{\mathrm{B}} T_{e\mathrm{h}\perp})}{B^2}, \tag{2.2}$$

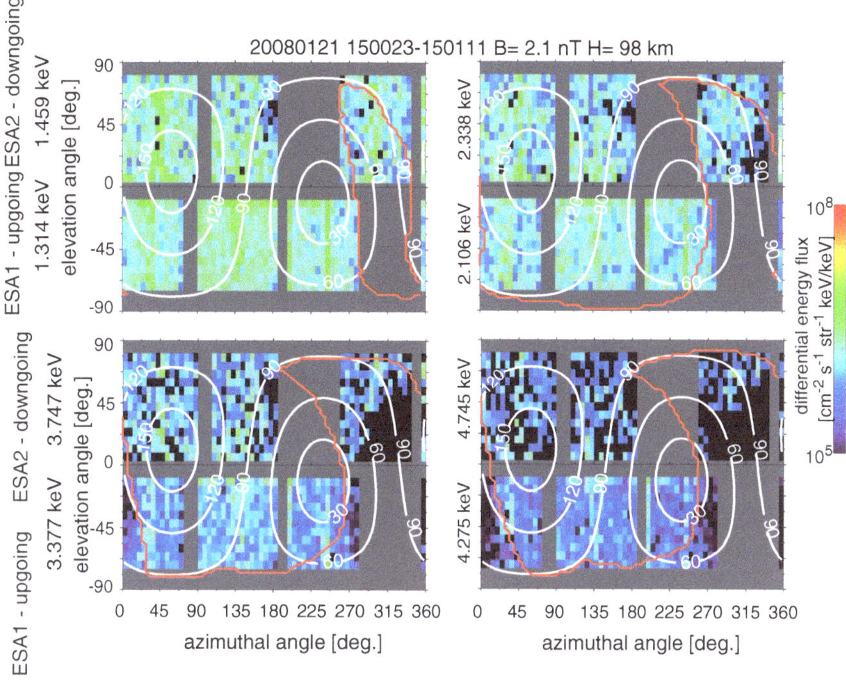

Fig. 2.13 Electron angular distribution obtained during 15:00:23–15:01:11 UT (48 s average) on 21 January 2008, when the Moon was in the Earth's magnetosphere, in the same format as in Fig. 2.8. The *red lines* indicate the forbidden regions for a uniform magnetic field and no electric field

where n_{eh} and T_{eh} are the density and temperature of hot electrons, respectively, and k_B is the Boltzmann constant. I assume that cold electrons do not contribute to the electron perpendicular pressure; i.e., $n_{ec}k_B T_{ec\perp} = 0$. If the magnetic field is oriented parallel to a plane solid surface ($+X$ direction), as illustrated in Fig. 2.2, the electron diamagnetic current flows parallel to the lunar surface ($+Y$ direction), since $\nabla p_{e\perp}$ directs away from the lunar surface ($-Z$ direction). In terms of the electron VDF, there are more electrons with negative than with positive Y velocity. These with positive Y velocity are partially lost because of the gyro-loss effect. Therefore, the $-Y$-directed bulk electron velocity, $\mathbf{V}_e$, produces an electron current, $\mathbf{J}_e$, in the $+Y$ direction.

Figure 2.16 shows a cross section of a modeled hot-electron VDF using the coordinate system defined in Fig. 2.2. I assume that the ambient electrons' velocity distribution is Maxwellian, with a density of $0.1\,\mathrm{cm}^{-3}$ and a temperature of 400 eV. I derive the forbidden region at an altitude of 50 km in the presence of a magnetic field of $\sim$3.2 nT. I can obtain the same result, no matter whether I use Eq. (2.1) or the particle-trace calculation, because the lunar surface is assumed to be planar while the magnetic field is assumed to be parallel to the lunar surface. The boundary of the forbidden region in the (v_y, v_z) plane becomes a parabola, as derived from Eq. (2.1);

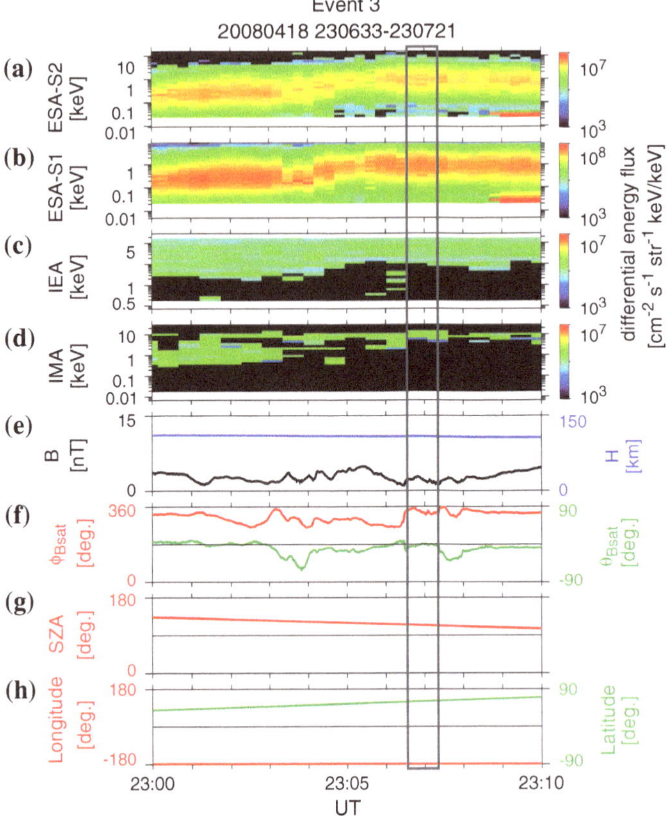

Fig. 2.14 Time-series data from Kaguya for Event 3 in the same format as in Fig. 2.9. The *gray box* indicates the time period of the gyro-loss event shown in Fig. 2.12

the gyro-loss condition $r_L \geq r_c$ yields $v_z^2 + (2eBH/m_e)v_y - (eBH/m_e)^2 \geq 0$, where m_e is the electron mass and e the elementary charge. Figure 2.16 shows that the bulk velocity of hot electrons $\mathbf{V}_{eh}$ has a $-Y$ component because a large part of the region for $v_y > 0$ is lost. By calculating the moments of this modeled electron VDF, n_{eh}, $\mathbf{V}_{eh}$, and $\mathbf{J}_e$ can be derived.

The distribution of n_{eh} calculated for each altitude is shown by black lines in the top left-hand panel of Fig. 2.17, V_{eh} is shown in the top right-hand panel, and J_e in the bottom left-hand panel. I calculated the moments for the energy range from -10 to $+10$ keV below an altitude of 150 km for 1 km steps. Here, $V_{eh} = |V_{ehy}|$ and $J_e = J_{ey}$, because the modeled electron VDF is symmetric relative to the (v_y, v_x) and (v_y, v_z) planes. The current layer with a peak at around 20 km is present below an altitude of 100 km, where the hot-electron density decreases and the corresponding velocity increases. The distribution of J_e depends on the parameters of the calculation; i.e., the ambient electron density, n_{e0}, the ambient

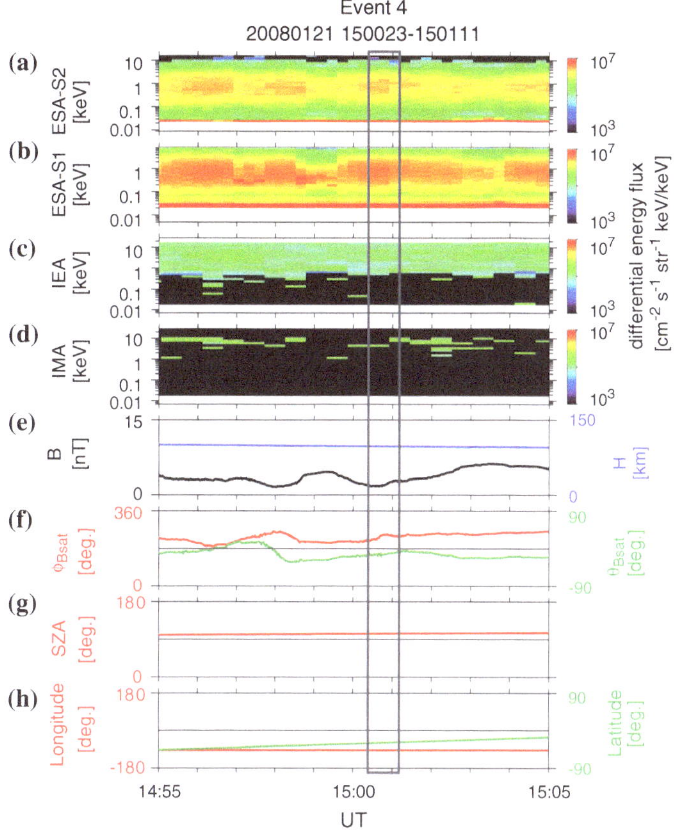

Fig. 2.15 Time-series data from Kaguya for Event 4 in the same format as in Fig. 2.9. The *gray box* indicates the time period of the gyro-loss event shown in Fig. 2.13

electron temperature, T_{e0}, and the ambient magnetic-field intensity, B_0. The vertical scale length of the current layer, L, depends on the electron gyrodiameter; i.e., $L \propto \sqrt{T_{e0}}/B_0$. Note that the average speed of the ambient electrons is proportional to $\sqrt{T_{e0}}$. If I define $L \equiv 100$ km based on Fig. 2.17, with $T_{e0} = 400$ eV and $B_0 \sim 3.2$ nT, I get

$$L \sim 16\frac{\sqrt{T_{e0}}}{B_0} \quad \text{km}, \tag{2.3}$$

with T_{e0} expressed in units of eV and B_0 in nT. Since the intensity of the current $J_e \propto T_{e0}n_{e0}/B_0L$—Eq. (2.2)—I obtain $J_e \propto n_{e0}\sqrt{T_{e0}}$.

The diamagnetic current produces a magnetic field that reduces the ambient magnetic field above and enhances the magnetic field below the current layer (see Fig. 2.2). Using Ampère's law, an infinite current sheet with a current per unit length I produces a magnetic field $B = \mu_0 I/2$. The magnetic field associated with the diamagnetic current is derived from Eq. (2.2) as

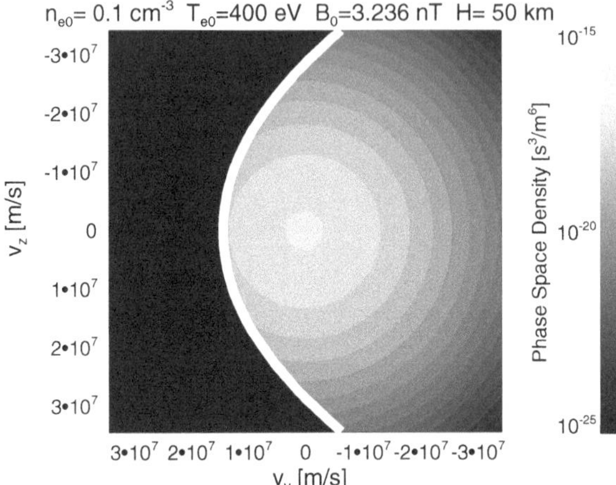

Fig. 2.16 Cross section of the modeled electron VDF at an altitude of 50 km with an ambient magnetic field of $B = 3.236\,\text{nT}$. The *white thick line* indicates the boundary of the forbidden region

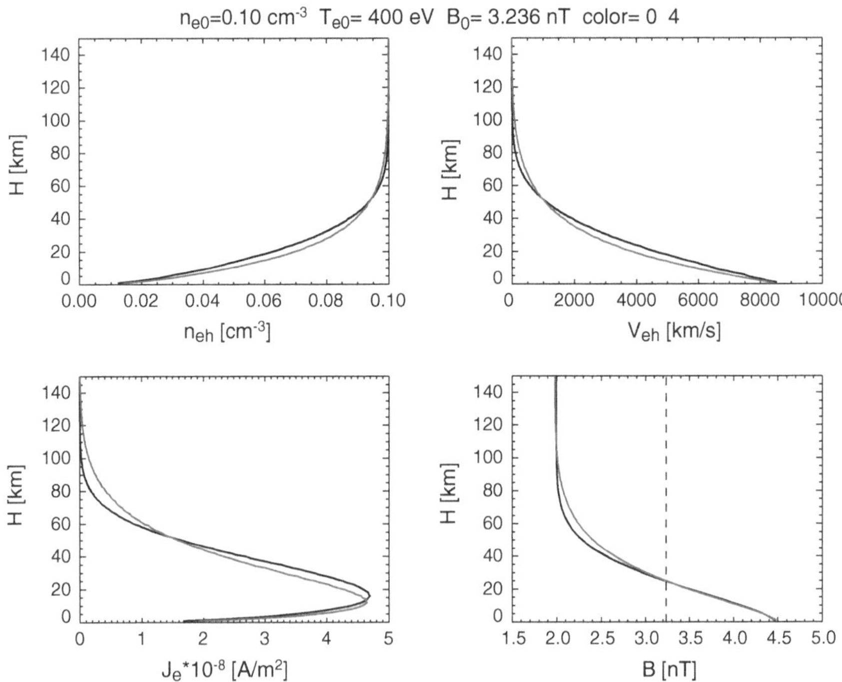

Fig. 2.17 Calculated electron moments in the diamagnetic-current model (see text for details). Iterated results are represented by the *gray lines*. The *vertical dashed line* in the *bottom right-hand panel* indicates the assumed ambient magnetic field strength of 3.236 nT

$$B_{\rm D} = \frac{\mu_0 I}{2} = \frac{\mu_0}{2}\frac{n_{e0}k_{\rm B}T_{e0}}{B_0 L}L \sim 0.1\frac{n_{e0}T_{e0}}{B_0} \quad \text{nT}, \tag{2.4}$$

with n_{e0} expressed in units of cm^{-3}, T_{e0} in eV, and B_0 in nT. For $n_{e0} = 0.1\,\text{cm}^{-3}$, $T_{e0} = 400\,\text{eV}$, and $B_0 \sim 3.2\,\text{nT}$, I obtain $B_{\rm D} \sim 1.2\,\text{nT}$.

Equation (2.4) expresses the magnetic-field strength outside the current layer produced by the diamagnetic current. The magnetic field produced by the diamagnetic current at an altitude H in the current layer is derived by adding the magnetic field produced by the currents above and below H,

$$B_{\rm D}(H) = -\int_0^H \frac{\mu_0 J_e(H')}{2}{\rm d}H' + \int_H^\infty \frac{\mu_0 J_e(H')}{2}{\rm d}H'. \tag{2.5}$$

The black line in the bottom right-hand panel of Fig. 2.17 shows the magnetic-field intensity for each altitude, derived as $B_0 + B_{\rm D}(H)$. I perform the integration over an interval of $[H, 150]$ km instead of $[H, \infty]$ in the second term of Eq. (2.5), since J_e above 150 km is much smaller than that below 150 km. As expected from Eq. (2.4), $B \sim 2.0\,\text{nT}$ above 100 km and $B \sim 4.5\,\text{nT}$ near the lunar surface in Fig. 2.17. In the current layer (0–100 km), the magnetic-field intensity increases gradually as H decreases.

Once this magnetic configuration is constructed, the electron motion and the forbidden regions in the electron VDFs are modified by the nonuniform magnetic field. I perform two-dimensional particle-trace calculations in the presence of a nonuniform magnetic field, linearly interpolated from $B_0 + B_{\rm D}(H)$, as indicated by the black line in the bottom right-hand panel of Fig. 2.17. I derived the forbidden regions for each altitude step. When I launch electrons from lower altitudes, I need more accuracy for the trajectory calculation to derive the correct forbidden regions. I trace electrons until they either strike the lunar surface or gyrate one cycle, with a time step of $T_{ce}/50$ if $H > 16\,\text{km}$ and $T_{ce}/(800/H)$ if $H \leq 16\,\text{km}$. After derivation of the forbidden regions, I calculate the moments of the electron VDF once again. I then obtain a new magnetic-field distribution by integrating Eq. (2.5). The gray lines in Fig. 2.17 show the results, iterated four times using this method. These results converge in this case as the calculation is iterated.

The magnetic-field distribution derived above can be applied to the 3D particle-trace calculation as long as the approximation of a plane lunar surface is valid; i.e., $H \ll R_{\rm L}$. The dashed lines in Fig. 2.18 show the forbidden regions derived from the 3D particle-trace calculation in the presence of a nonuniform magnetic field produced by the electron diamagnetic current. I assume that the magnetic-field intensity distribution depends on H, which I derive by iterating four times (as indicated by the gray lines in Fig. 2.17). I trace electrons until they either strike the lunar surface or gyrate one cycle, adopting a time step of $T_{ce}/50$. In this case, $B \sim 2.0\,\text{nT}$ at 100 km and the magnetic field is enhanced as H decreases. Some of the electrons are magnetically deflected before they strike the lunar surface, resulting in smaller forbidden regions than for a uniform magnetic field. As expected from Eqs. (2.3)

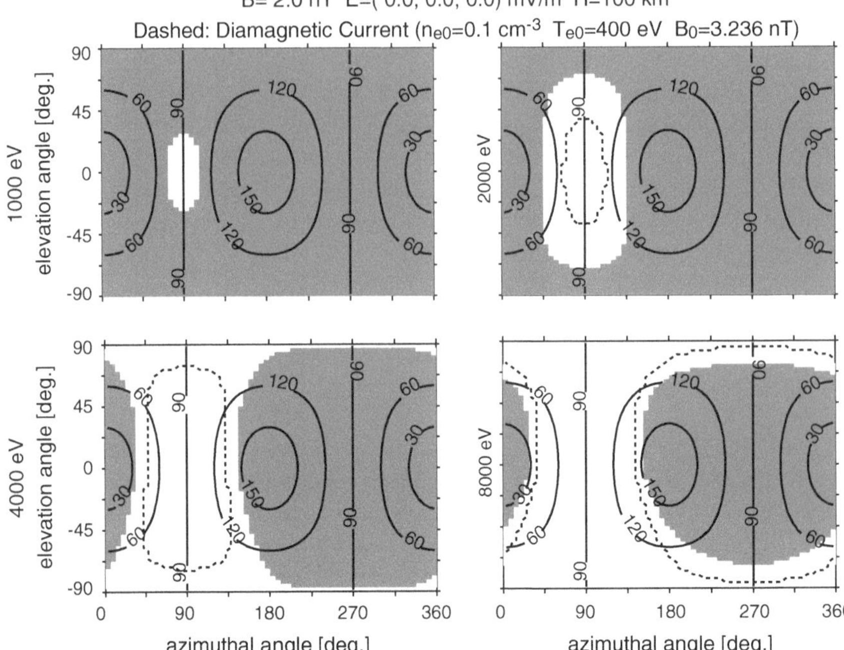

Fig. 2.18 Forbidden regions in the electron VDF considering the diamagnetic-current effect, in the same format as in Fig. 2.5. The *white regions* indicate the same forbidden regions as in Fig. 2.5. The *dashed lines* indicate those taking into account the nonuniform magnetic field produced by the diamagnetic current near the lunar surface

and (2.4), our calculation results of the modified forbidden regions including this diamagnetic-current effect depend on the ambient electron density, temperature, and magnetic-field strength.

I assumed that loss of hot electrons with large gyroradii, hence with large $v_\perp$, is compensated by cold electrons with $v_\perp = 0$, and therefore, $n_{ec}k_B T_{ec\perp} = 0$. If hot electrons with $v_\perp \neq 0$ intrude along the field line, they will contribute to the electron perpendicular pressure. However, they have smaller $v_\perp$ than those of the lost electrons because the gyroradii of the intruding electrons must be smaller than the critical gyroradius, otherwise they will strike the lunar surface before they reach the loss region. Since the number of these intruding electrons does not exceed that of lost electrons to satisfy the charge neutrality condition, the loss of electron perpendicular pressure cannot be completely compensated. In other words, replacement of hot electrons by "colder" electrons (not necessarily $n_{ec}k_B T_{ec\perp} = 0$) results in loss of electron perpendicular pressure. Therefore, the gradient of electron perpendicular pressure is left to some extent, and the diamagnetic current does not vanish completely by hot electron intrusion, although it might be weakened compared to our calculation. I also have to be aware that our estimation of the magnetic-field variations produced by the diamagnetic-current layer might be overestimated, because I do not consider the spherical lunar surface in our calculation of the current distribution.

2.3.3 Lunar Surface Charging

If the lunar surface is charged negatively, electrons are thought to be reflected by an electric field near the lunar surface (see Fig. 2.19). The electric field produced by the surface potential can be shielded within a few Debye lengths (<1 km), which is much smaller than the electron gyroradius considered here (>10 km) [10]. Therefore, I assume that the electrons are nonmagnetized on the scale length of the surface potential. If the electric field is assumed to be radial with respect to the Moon, from the relationship between the radial velocity component of the electrons at the lunar surface $\mathbf{v}_{\mathrm{rad}}$ and the lunar surface's electrostatic potential U_{M} (<0 V), the reflection condition of the electrons is given by

$$v_{\mathrm{rad}} < \sqrt{\frac{-2eU_{\mathrm{M}}}{m_e}}. \tag{2.6}$$

I now derive the forbidden regions in the electron VDFs by particle-trace calculation, taking this condition into account.

The dashed lines in Fig. 2.20 indicate the forbidden regions for $U_{\mathrm{M}} = -500$ V in Eq. (2.6), and the thin lines indicate the results for $U_{\mathrm{M}} = -1{,}000$ V. Since electrons with radial velocity components that satisfy the inequality (2.6) are reflected at the lunar surface, these forbidden regions are smaller than those in the absence of negative lunar-surface charging (see the white regions in Fig. 2.20). The forbidden regions with energy of 1 keV for $U_{\mathrm{M}} =$ 500 and $-1{,}000$ V disappear (top left-hand panel). On

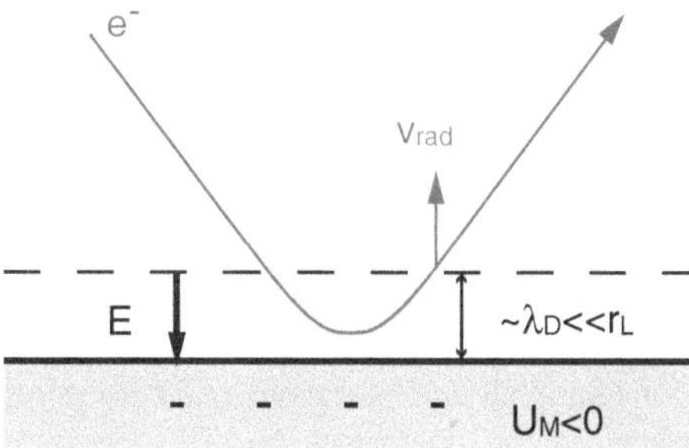

Fig. 2.19 Schematic illustration of an electron trajectory near the negatively charged lunar surface; $\mathbf{v}_{\mathrm{rad}}$ is the radial velocity component to the Moon, and the electric field **E** produced by the negative surface electrostatic potential U_{M} is shielded within a few Debye lengths, λ_{D}, which is much smaller than the electron gyroradius, r_L

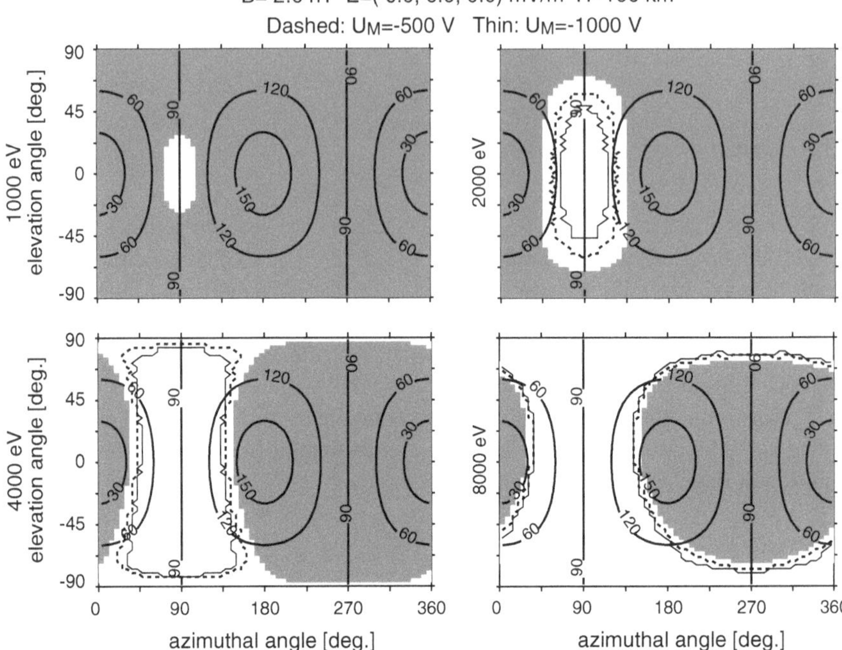

Fig. 2.20 Forbidden regions in the electron VDF considering the lunar-surface charging, in the same format as in Fig. 2.5. The *white regions* indicate the same forbidden regions as in Fig. 2.5. The *dashed lines* indicate those for an electrostatic potential at the lunar surface of −500 V, and the *thin lines* indicate those for a surface potential of −1,000 V

the other hand, the forbidden regions with an energy of 8 keV for $U_{\mathrm{M}} = -500$ and −1,000 V are not very different from those generated in the absence of negative lunar-surface charging. At higher energies and large v_{rad}, fewer electrons satisfy inequality (2.6). For larger lunar-surface potentials, the forbidden regions become smaller, since more electrons satisfy inequality (2.6).

2.3.4 Perpendicular Electric Fields

If an electric field component exists perpendicularly to the magnetic field and the scale length of the region characterized by the electric field is longer than the electron gyrodiameter, electrons will drift with a $\mathbf{E} \times \mathbf{B}$ drift velocity $v_{E\times B} = E_{\perp}/B$. The forbidden regions in the electron VDFs will be modified, since the electron motion is different from that in the absence of electric fields. I perform particle-trace calculations and derive the forbidden regions, taking into account four cases; i.e., $E_y < 0$, $E_y > 0$, $E_z < 0$, and $E_z > 0$, all in the presence of uniform magnetic fields oriented

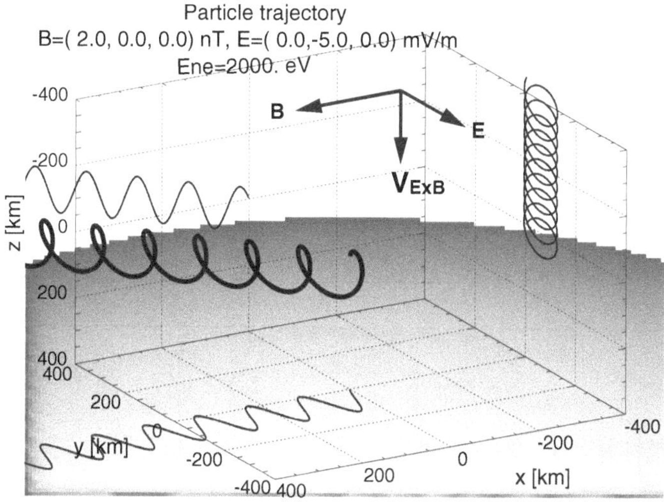

Fig. 2.21 Sample electron trajectories in the presence of a uniform electric field of $E_y = -5\,\mathrm{mV/m}$

along the X axis. I trace electrons until they either strike the lunar surface or gyrate twenty cycles ($t \geq -20\,T_{ce}$). I assume a relatively strong electric field of 5 mV/m in each case to generate obvious effects for further study.

For $E_y < 0$, electrons drift in the $+Z$ direction (towards the lunar surface) as shown in Fig. 2.21. In this case, the forbidden regions are indicated by the red lines in Fig. 2.22. The top left-hand panel, for an energy of 1 keV, shows no forbidden region for $E_y = -5\,\mathrm{mV/m}$. The other panels show that the forbidden regions are smaller than those found in the absence of an electric field and they are displaced towards the bottom in all panels. This occurs because the effective height H' in the guiding center's rest frame becomes larger than H because of the $\mathbf{E} \times \mathbf{B}$ drift and H' depends on the gyrophase of the electrons that enter the sensors.

For $E_y > 0$, electrons drift in the $-Z$ direction (away from the lunar surface) and the forbidden regions are shown by the orange lines in Fig. 2.22. These forbidden regions have rather different distributions from those for the other cases, especially for the lower-energy electrons. Note that the regions around pitch angles $\alpha = 90^\circ$ correspond to the forbidden regions, for example in the top left-hand panel, the region around $60 < \alpha < 120^\circ$ indicates the forbidden region. Figure 2.23 shows the trajectories of an electron outside and in the forbidden region by the light blue and purple traces, respectively (the initial angles are indicated by the circles and arrows of the same colors in the top right-hand panel of Fig. 2.22). Electrons with pitch angles around 90° have small velocity components parallel to the magnetic field, so they drift towards the lunar surface under reverse tracking and strike the lunar surface, as shown by the purple trace. On the other hand, electrons with large velocity components parallel to the magnetic field do not strike the lunar surface,

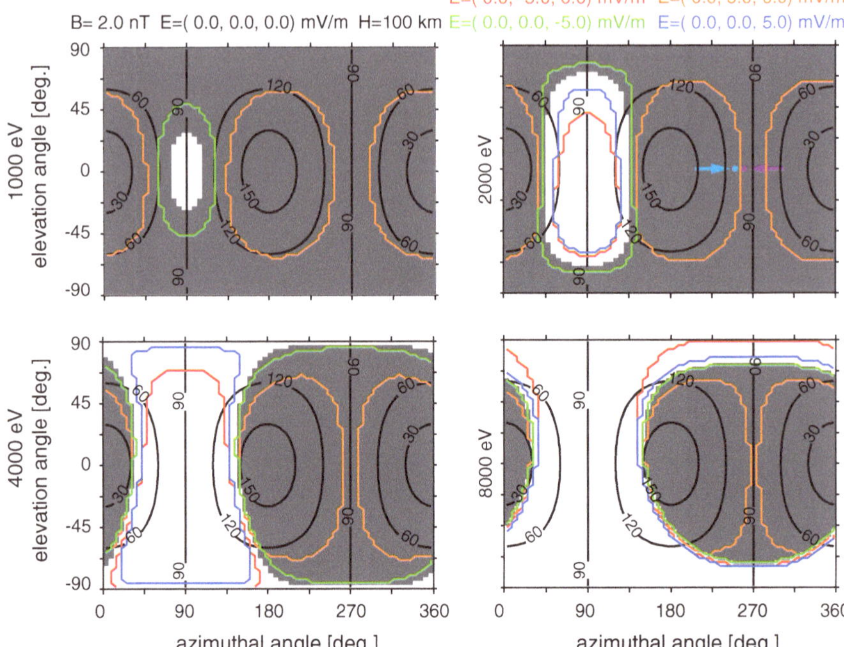

Fig. 2.22 Forbidden regions in the electron VDF for a uniform magnetic field ($B_x = 2\,\mathrm{nT}$) and uniform electric fields perpendicular to the magnetic field, in the same format as in Fig. 2.5. The *white regions* indicate the forbidden regions in the absence of electric fields. The *red lines* indicate those for $E_y = -5\,\mathrm{mV/m}$, the *orange lines* indicate those for $E_y = 5\,\mathrm{mV/m}$, the *green lines* indicate those for $E_z = -5\,\mathrm{mV/m}$, and the *blue lines* indicate those for $E_z = 5\,\mathrm{mV/m}$. The *light blue circle* and *arrow* in the *top right-hand panel* indicate the (250°, 0°) point, which corresponds to the *light blue trajectory* in Fig. 2.23 for $E_y = 5\,\mathrm{mV/m}$, and the *purple circle* and *arrow* indicate the (260°, 0°) point, corresponding to the *purple trajectory* in Fig. 2.23

because I assume that the Moon is a sphere, as shown by the light blue trace in Fig. 2.23. The forbidden regions for higher energies exhibit similar distributions to those in the absence of an electric field, because the $\mathbf{E} \times \mathbf{B}$ drift becomes less effective.

For $E_z < 0$ and $E_z > 0$, the forbidden regions are indicated by the green and blue lines in Fig. 2.22, respectively. The forbidden regions for $E_z < 0$ (indicated by the green lines) become larger and those for $E_z > 0$ (indicated by the blue lines) become smaller than those in the absence of electric fields. This can be explained as follows. If an electron enters the sensor with a velocity $\mathbf{v}_\perp = (0, v_y, v_z)$, as illustrated in Fig. 2.3. Equation (2.1) becomes

$$r_\mathrm{c} = \frac{H}{1 - \cos\psi} = \frac{H}{1 + v_y/v_\perp}, \tag{2.7}$$

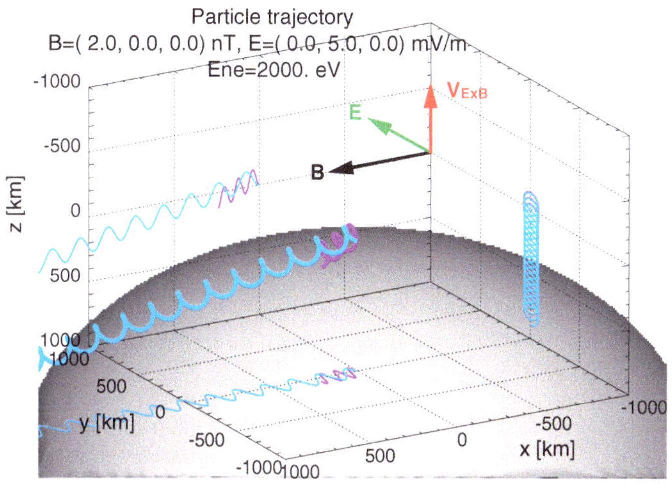

Fig. 2.23 Sample electron trajectories for $E_y = 5\,\text{mV/m}$; the *light blue* trace shows the trajectory launched with an initial azimuthal angle of 250° and an elevation angle of 0°, the *purple trace* shows the trajectory with an azimuthal angle of 260° and an elevation angle of 0°

where $v_\perp = |\mathbf{v}_\perp|$. Using this relation, the condition that the electron strike the lunar surface, $r_L \geq r_c$, is converted to

$$\frac{m_e v_\perp}{eB} \geq \frac{H}{1 + v_y/v_\perp} \tag{2.8}$$

$$v_\perp + v_y - \frac{eBH}{m_e} \geq 0. \tag{2.9}$$

For $E_z < 0$ and $E_z > 0$, an electron drifts into the $-Y$ and $+Y$ directions, respectively, and the velocity of the electron in the guiding center's rest frame is given by $\mathbf{v}'_\perp = \mathbf{v}_\perp - \mathbf{v}_{E\times B} = (0, v_y - E_z/B, v_z)$. Hence, inequality (2.9) is modified to

$$v'_\perp + v_y - \frac{E_z}{B} - \frac{eBH}{m_e} \geq 0. \tag{2.10}$$

Since (left-hand side [LHS] of inequality (2.9)) − (LHS of inequality (2.10)) = $v'_\perp - v_\perp - E_z/B$, and using triangle inequalities, $v'_\perp + |E_z/B| \geq v_\perp$ and $v_\perp + |E_z/B| \geq v'_\perp$, I obtain the following relationships: if $E_z < 0$ (LHS of inequality (2.10)) $\geq$ (LHS of inequality (2.9)), and if $E_z > 0$ (LHS of inequality (2.10)) $\leq$ (LHS of inequality (2.9)). Therefore, for $E_z < 0$, more electrons satisfy inequality (2.10) and the forbidden regions become large. In contrast, fewer electrons satisfy equality (2.10) and the forbidden regions become small for $E_z > 0$.

Finally, I summarize the perpendicular electric-field influence on the forbidden regions in the electron VDFs caused by the gyro-loss effect. For $E_y < 0$, the forbidden regions become smaller and they shift in the $-v_z$ direction. For $E_y > 0$, the forbidden regions appear like the regions with pitch angles of approximately 90°. For $E_z < 0$, the forbidden regions become larger. For $E_z > 0$, the forbidden regions become smaller. In all cases, perpendicular electric-field effects appear less clearly in the forbidden regions for higher energies.

2.3.5 Discussion

Here I discuss the electrons' interaction with the lunar surface, as well as with the lunar plasma environment, by comparing the observed non-gyrotropic empty regions with the forbidden regions modified by taking into account the mechanisms that I have discussed.

During Event 1, the observed electron density and temperature were relatively low and the magnetic-field intensity was fairly strong ($0.10\,\mathrm{cm}^{-3}$, 88 eV, and 4.6 nT). Therefore, the diamagnetic-current effect is not significant in this case. If I assume $n_{e0} = 0.10\,\mathrm{cm}^{-3}$, $T_{e0} = 88\,\mathrm{eV}$, and $B_0 = 4.6\,\mathrm{nT}$, I obtain $B_\mathrm{D} \sim 0.2\,\mathrm{nT}$ using Eq. (2.4). Using these observed values as ambient-plasma parameters is, in fact, not appropriate, since Kaguya was thought to be located in the diamagnetic-current layer; using Eq. (2.3), I estimate $L \sim 33\,\mathrm{km} > H = 12\,\mathrm{km}$. However, I can at least estimate the order of magnitude of the magnetic-field strength produced by the diamagnetic-current system, and this magnetic field seems to be too weak to modify the electron motion.

The red dashed lines in Fig. 2.24 indicate the modified forbidden regions for $U_\mathrm{M} = -50\,\mathrm{V}$ in Eq. (2.6). The forbidden regions become small, because some of the electrons are reflected by the lunar surface's electrostatic potential. The red dashed lines seem to correspond to the inner edges of the observed empty regions, suggesting that the electrons in the regions between the red solid and dashed lines were reflected because of lunar-surface charging. The surface potential of $-50\,\mathrm{V}$ is quite reasonable compared with the observed electron temperature of 88 eV and previous Lunar Prospector observations, which show potentials of -150 to 0 V in the magnetotail lobes [13]. In addition, although Kaguya were not able to observe an upward-going electron beam from the lunar surface during Event 1 (because the magnetic-field line passing through the spacecraft was not connected to the Moon), electron beams with energies of $\sim$50 eV were observed by Kaguya just before and after the time interval of Event 1 as shown by the black arrows in Fig. 2.9. Therefore, despite the fact that I cannot completely exclude the possibilities of perpendicular electric-field and magnetic-anomalies-related effects, lunar-surface charging is thought to be the most plausible mechanism to produce the slight differences between the observed empty and forbidden regions which assume a uniform magnetic field and the absence of any electric field.

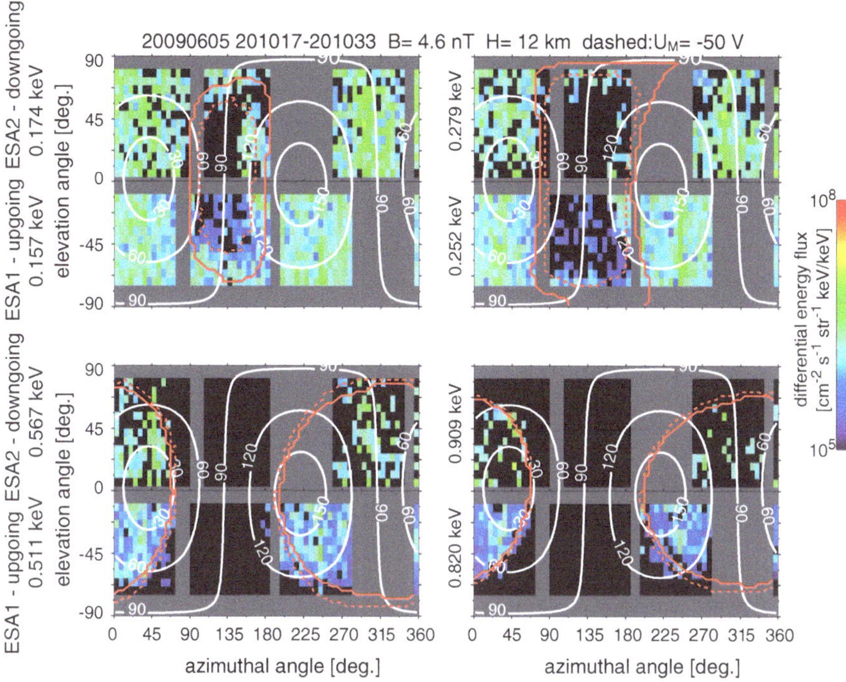

Fig. 2.24 Electron angular distribution for Event 2. All panels are identical to those shown in Fig. 2.8 except that the *red dashed lines* indicate the modified forbidden regions for an electrostatic potential of the lunar surface of −50 V

The Moon was located in the solar wind during Event 2, and lunar-surface charging would not contribute significantly, because Kaguya was located on the dayside of the Moon (solar zenith angle 39°), where the lunar surface is usually charged positively [13]. When Kaguya is located above the current layer or near its top edge, most electrons are not lost, except for higher-energy electrons, as shown in Fig. 2.17. Therefore, I can use the observed density and electron temperature as approximations of the ambient-plasma parameters to estimate the diamagnetic-current effect as described in Sect. 2.3.2. During Event 2, Kaguya was presumably located above the current layer; if I assume $n_{e0} = 3.20\,\text{cm}^{-3}$, $T_{e0} = 21\,\text{eV}$, and $B_0 = 6.2\,\text{nT}$, I obtain $L \sim 12\,\text{km}$ from Eq. (2.3) and $B_D \sim 1.1\,\text{nT}$ from Eq. (2.4). Therefore, $B \sim 5.1\,\text{nT}$ at $H = 14\,\text{km}$, which is consistent with the observed magnetic-field intensity. The modified forbidden regions in the presence of the diamagnetic-current effect are indicated by the orange lines in Fig. 2.25. In addition, I can estimate the perpendicular electric-field effect by deriving the solar-wind motional electric field $\mathbf{E} = -\mathbf{V} \times \mathbf{B}$ from the data obtained by IEA and LMAG. The yellow lines in Fig. 2.25 indicate the modified forbidden regions, taking into account the solar-wind motional electric-field effect, while the green lines indicate the forbidden regions when both the diamagnetic-current and the perpendicular electric-field effects operate. The green

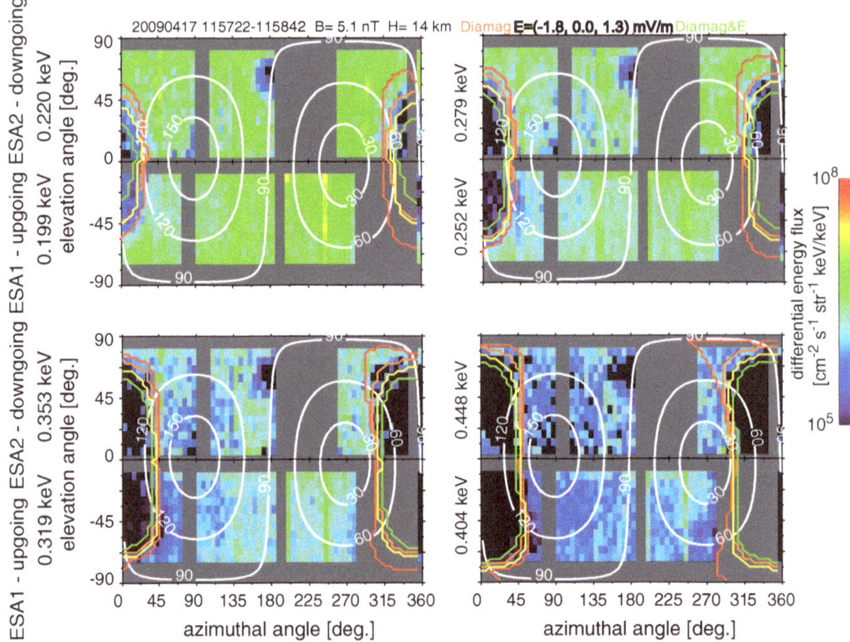

Fig. 2.25 Electron angular distribution for Event 2. All panels are identical to those shown in Fig. 2.10 except that the *orange lines* indicate the modified forbidden regions under the diamagnetic-current effect, the *yellow lines* indicate those affected by a perpendicular electric field, and the *green lines* indicate those due to both the diamagnetic-current effect and a perpendicular electric field. A solar-wind motional electric field of 2.3 mV/m is estimated from $-\mathbf{V} \times \mathbf{B}$ using the data obtained by IEA and LMAG

lines seem to be somewhat smaller than the observed empty regions, suggesting that one or both of these effects might be slightly overestimated. In any case, the orange and yellow lines are in general agreement with the observed empty regions, indicating that the gyro-loss effect can be affected by nonuniform magnetic fields caused by the diamagnetic-current system and/or the electron-drift motion because of the solar-wind motional electric field.

During Event 3, the Moon was located in the central plasma sheet and Kaguya was thought to be located near the top edge of the diamagnetic-current layer; if I assume $n_{e0} = 0.06\,\text{cm}^{-3}$, $T_{e0} = 446\,\text{eV}$, and $B_0 = 3.0\,\text{nT}$, I obtain $L \sim 113\,\text{km}$ from Eq. (2.3) and $B_\text{D} \sim 0.9\,\text{nT}$ from Eq. (2.4), resulting in $B \sim 2.1\,\text{nT}$ at $H = 109\,\text{km}$. Using the observed density and electron temperature as ambient-plasma parameters, the modified forbidden regions in the presence of the diamagnetic-current effect are derived as indicated by the orange solid lines in Fig. 2.26. Although the diamagnetic-current effect works substantially because of the relatively high electron temperature and the weak magnetic field, this effect does not seem to be sufficient to account for the electrons observed in the forbidden regions. From the data obtained by IEA and

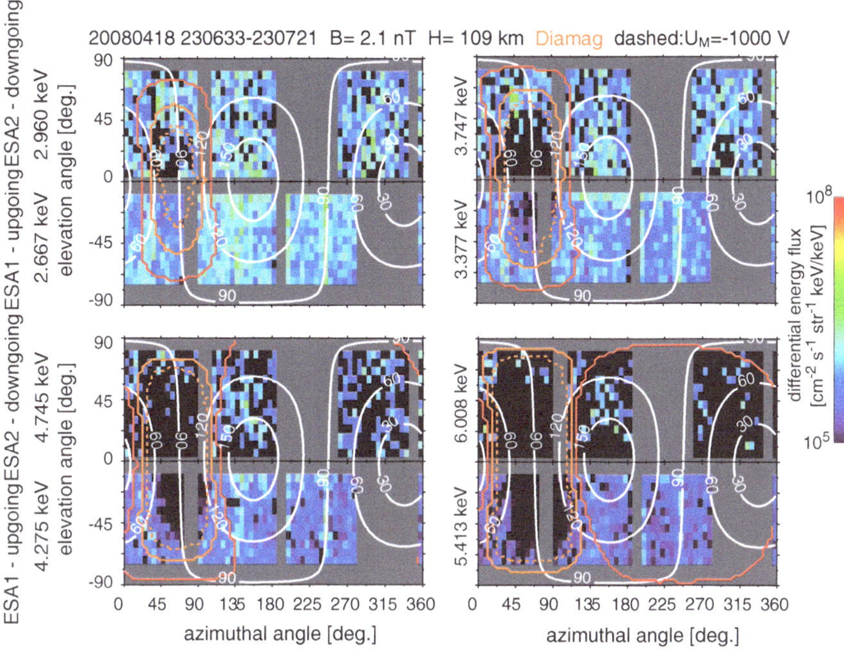

Fig. 2.26 Electron angular distribution for Event 3. All panels are identical to those shown in Fig. 2.12 except that the *orange solid lines* indicate the modified forbidden regions under the diamagnetic-current effect, and the *orange dashed lines* indicate those affected by the diamagnetic-current effect and an electrostatic potential of the lunar surface of −1,000 V

LMAG during Event 3, the convection electric field of the Earth's magnetosphere is estimated to be lower than 1.0 mV/m, which can hardly modify the forbidden regions with energies greater than a few keV (not shown in Fig. 2.26 because they almost coincide with the orange lines).

When the Moon is located in the terrestrial plasma sheet, it is known that the lunar-surface potentials can be up to ∼−1,000 V [13]. The modified forbidden regions that take into account the lunar surface-charging effect for $U_M = -1{,}000$ V in addition to the diamagnetic-current effect are indicated by the orange dashed lines in Fig. 2.26, which roughly correspond to the observed empty regions. A small fraction of the electron population is still found in the modified forbidden regions, but it seems that the diamagnetic-current effect and lunar-surface charging are the predominant mechanisms responsible for modifying the forbidden regions during Event 3.

Kaguya was presumably located near the top edge of the diamagnetic-current layer during Event 4; if I assume $n_{e0} = 0.07\,\text{cm}^{-3}$, $T_{e0} = 435$ eV, and $B_0 = 3.1$ nT, I obtain $L \sim 107$ km from Eq. (2.3) and $B_D \sim 1.0$ nT from Eq. (2.4). The orange solid lines in Fig. 2.27 indicate the modified forbidden regions, taking into account the diamagnetic-current effect using these values as ambient-plasma parameters, and the orange dashed lines indicate the forbidden regions in the presence of the

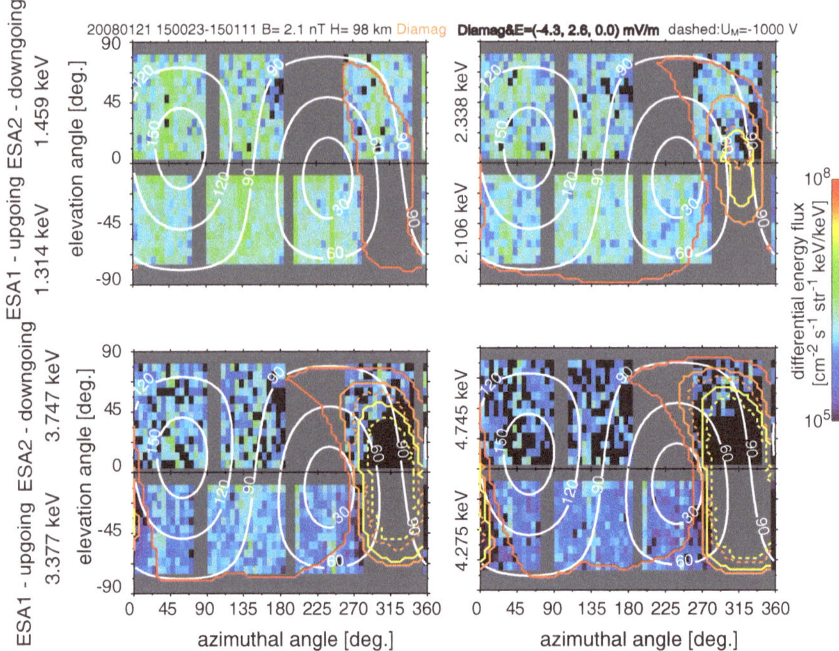

Fig. 2.27 Electron angular distribution for Event 4. All panels are identical to those shown in Fig. 2.13 except that the *orange solid lines* indicate the modified forbidden regions under the diamagnetic-current effect, the *orange dashed lines* indicate those affected by the diamagnetic-current effect and an electrostatic potential of the lunar surface of $-1{,}000$ V, the *yellow solid lines* indicate those due to the diamagnetic-current effect and a perpendicular electric field of 5 mV/m, and the *yellow dashed lines* indicate those under the diamagnetic-current effect, a perpendicular electric field of 5 mV/m, and a lunar-surface potential of $-1{,}000$ V

diamagnetic-current effect and a lunar-surface potential of $-1{,}000$ V. These modified forbidden regions under the influence of the diamagnetic-current effect and lunar-surface charging do not seem to be in agreement with the observed empty regions, especially not in the regions with positive elevation angles.

Finally, I take into account the perpendicular electric-field effect during Event 4. The estimated convection electric field of the Earth's magnetosphere (<0.9 mV/m) can hardly modify the forbidden regions with energies of a few keV. However, an electric field can become locally stronger than the convection electric field if the scale length of the region characterized by the strong electric field is shorter than the ion gyrodiameter. In this case, electrons execute $\mathbf{E} \times \mathbf{B}$ drift, but ions do not. Such small-scale electric fields can significantly modify the electron motion. The yellow solid lines in Fig. 2.27 indicate the modified forbidden regions under a perpendicular electric field of 5 mV/m and the diamagnetic-current effect, while the yellow dashed lines indicate those where I additionally assumed a surface potential of $-1{,}000$ V. I assume an electric field corresponding to $E_y < 0$ in Sect. 2.3.4, because the forbidden regions with positive elevation angles become smallest in this case. The yellow lines seem to be more consistent with the observed empty regions than the orange lines, suggesting that an electric field of at least 5 mV/m is needed to explain this event by the mechanisms discussed in this section. Even if this electric field exists, the mechanism responsible for the production of such a strong electric field is unknown at the present. Note that although this event was observed when Kaguya was located above the rather weak and/or small-scale crustal-field regions, the small-scale crustal-field effect cannot be ruled out.

The electron gyro-loss events observed in the Earth's magnetosphere and the solar wind have been analyzed by considering the three mechanisms that modify the forbidden regions in the electron VDFs. Each event shows that the characteristics of the non-gyrotropic electron VDFs associated with the gyro-loss effect depend significantly on the ambient plasma and the lunar-surface conditions. I selected the events observed near the weak and/or small-scale crustal-field regions in this section, but if nonuniform magnetic fields related to lunar magnetic anomalies are taken into account, the process responsible for producing non-gyrotropic electron VDFs will be even more complicated. In the next section, I will further present Kaguya observations of complex-shape "empty regions" presumably associated with small-scale crustal magnetic fields on the lunar surface.

Kaguya's low orbital altitudes of less than the electron gyrodiameter and the excellent angular resolutions of ESA-S1 and ESA-S2 enable clear detection of non-gyrotropic electron VDFs. We now know that non-gyrotropic electrons commonly exist near the lunar surface and that spacecraft orbiting at lower altitudes than ~100 km are able to detect them. The vicinity of the lunar surface would be one of the suitable environments for investigation of non-gyrotropic electrons, including their wave–particle interactions, which have rarely been studied observationally because of the difficulty of detecting non-gyrotropic electron VDFs.

Non-gyrotropic electrons near the lunar surface might have implications for surface electrostatic potentials and dust dynamics. Shadowing of the incoming flux depending on the ambient magnetic field direction as well as photoelectrons returned to the surface by crustal magnetic fields could alter the incident currents to the surface, and therefore, the equilibrium surface potential [33]. In addition to shadowing effect on the total incident flux to a smooth surface, surface topography might be important to local potential distribution. For example, it was suggested that topographic features such as mountains or crater rims could create local sunlit/shadow regions and local plasma void, leading to the formation of local electric fields and the acceleration of charged dust grains [10]. Highly angular-dependent hot electrons produced by the gyro-loss effect could also influence the surface potential distribution according to the surface topography, causing local electric fields and charged dust motion. This effect would be significant especially in the plasma sheet, where large incident flux of hot electrons is expected.

While the loss of hot electrons in the plasma sheet including its angular dependence was observed on the lunar surface [33], magnetic-field variations associated with the diamagnetic-current layer near the lunar surface have not been reported thus far. Even if indeed the near-surface current layer exists, there seems to be some difficulty to find the clear signatures of the diamagnetic-current-related fields. Magnetic shadowing occurred infrequently at the Apollo sites located at relatively low latitudes and low longitudes, because magnetic fields in the Earth's magnetosphere at lunar distance are typically B_x-dominant and they rarely become parallel to the lunar surface there [32]. In the solar wind, on the other hand, the interaction of crustal magnetic fields with the solar-wind plasma would be predominantly observed on the surface [9].

Single spacecraft measurement cannot distinguish the ambient field $\mathbf{B}_0$ and the diamagnetic-current field $\mathbf{B}_D$. In order to detect $\mathbf{B}_D$ by a single probe, it needs to move along the gradient of perpendicular pressure (across the current layer) faster than $\mathbf{B}_0$ changes. Such a trajectory seems to be difficult for the electron-perpendicular-pressure gradient near the lunar surface, since the probe needs to sweep the altitudinal structure faster than $\mathbf{B}_0$ changes. In contrast, it is quite possible for the pressure gradient across the wake boundary, where various spacecraft have observed diamagnetic-current-related fields. It seems possible to find the signatures of $\mathbf{B}_D$ near the lunar surface using two probes, but very careful investigation would be needed in order to identify $\mathbf{B}_D$, because the ambient magnetic field is frequently turbulent in the plasma sheet or in the solar wind.

It would be a real challenge to understand the current systems near the Moon comprehensively, including the diamagnetic current layer near the lunar surface. If the diamagnetic current actually exists near the lunar surface, it might be connected to the diamagnetic current system near the wake boundary in the solar wind, and perhaps to some kind of currents associated with magnetic anomalies, which might form a miniature magnetosphere [22]. Further analysis of data obtained by recent missions such as Kaguya or ARTEMIS, as well as careful reanalysis of data in the Apollo era, will be necessary to understand the lunar plasma environment from the viewpoint of current systems.

2.4 Small-Scale Magnetic Fields on the Lunar Surface Inferred from Plasma Sheet Electrons

This section deals with crustal magnetic field effects on the plasma sheet electron motion near the lunar surface, which are neglected in Sect. 2.3. The motion of hot electrons with gyroradii larger than the scale size of the crustal magnetization no longer conserves the first adiabatic invariant (the magnetic moment of the electron). Since their trajectories can become very complex, I conduct particle trace calculations which include 3D lunar magnetic field data to investigate the electron motion in a realistic magnetic field configuration. In turn, the magnetic field distribution on the lunar surface can be inferred from the electron and magnetometer measurements in the terrestrial plasma sheet.

As introduced in Sect. 2.1.2, the difficulty in mapping kilometer-scale magnetic fields hinders our ability to interpret the surface distribution of crustal magnetic fields in a comprehensive way. In this section, I present a new technique to map the small-scale magnetic fields from electron and magnetic field observations in the terrestrial magnetotail, using data from instruments on the Kaguya spacecraft. In contrast to the ER technique, I use non-adiabatic scattering of plasma sheet electrons with energies >1 keV. The combination of the electron observations by Kaguya MAP–PACE and particle-tracing calculations using lunar magnetic field data synthesized from the Kaguya MAP–LMAG measurements allows us to extract valuable information about the small-scale crustal magnetic fields of the Moon. Hereafter, I use "large-" and "small-"scale magnetic fields to describe crustal magnetic fields with spatial scales larger and smaller, respectively, than the lowest orbital altitude of Kaguya at which the magnetometer measurements are used for producing the synthesized lunar magnetic field data (i.e., ~20 km).

2.4.1 Mapping Small-Scale Magnetic Fields

When the Moon is located in the terrestrial plasma sheet, the lunar surface is exposed to hot electrons. These relatively high-energy electrons are usually absorbed when they strike the lunar surface. This results in a partial loss in the electron velocity distribution function. However, Kaguya sometimes observed high-energy electrons with energies >1 keV traveling upward from a particular local area on the lunar surface. Figure 2.28 shows electron angular distributions from one of these observations at a low altitude of 23 km. The observed empty regions are highly modified from the "uniform-magnetic-field" forbidden regions (cf. the red lines), suggesting a non-adiabatic nature of the electron motion. These electron VDFs exhibit more complicated shapes of the empty regions compared to the electron VDFs shown in Sect. 2.3. Such complicated shapes of empty regions are not expected by the modification mechanisms considered in Sect. 2.3. Figure 2.29 shows electron trajectories derived from these electron distributions. Here I trace back the electrons from the spacecraft in a spatially uniform magnetic field. These electrons seem to come from a local area on the lunar surface within a single gyration, implying non-adiabatic scattering by the crustal magnetic fields.

I now investigate the effect of the large-scale magnetic field on high-energy electron trajectories by introducing three-dimensional lunar magnetic field data synthesized from magnetometer measurements into our particle-tracing calculations. I use the Equivalent Pole Reduction (EPR) method, one of the equivalent source models for mapping large-scale crustal magnetic fields [38, 39]. This method solves the inverse problem of a surface density distribution of equivalent monopoles ("magnetic charges") to fit the magnetic field measurements. The three components of the lunar magnetic fields are calculated at arbitrary positions above the surface from positive and negative magnetic charges distributed on the surface. I applied the EPR method to the LMAG measurements obtained over 12 consecutive orbits on 7 May 2009 at low altitudes of ~20–30 km (Fig. 2.30a) in the terrestrial magnetotail (plasma densities <0.3 cm^{-3} derived from the moment calculation). Figure 2.30b shows the

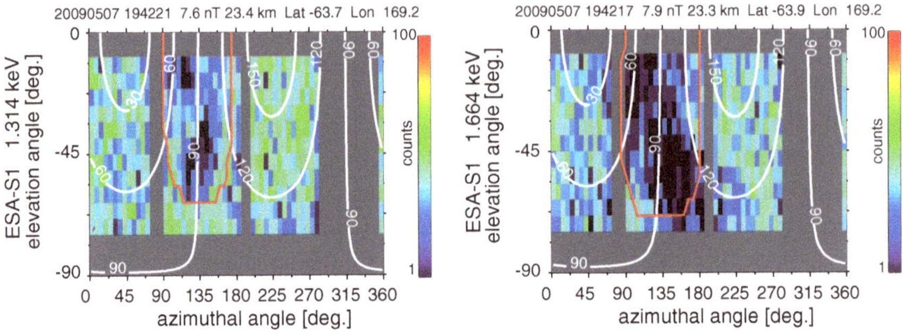

Fig. 2.28 Electron angular distributions from ESA-S1 onboard Kaguya for (*left*) 1.314 keV and (*right*) 1.664 keV from one energy scan at 19:42:11–19:42:27 UT on 7 May 2009. Angles with little or no sensitivity are indicated in *gray*. The *white contours* represent the pitch angles. The *red lines* indicate the regions where the electron absorption by the lunar surface is expected from particle-trace calculations in a uniform magnetic field

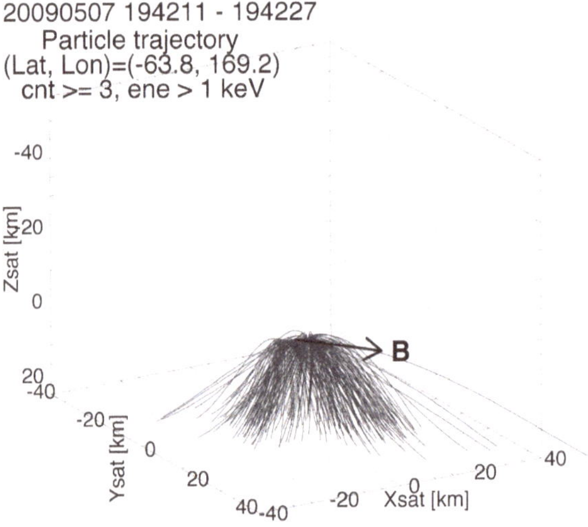

Fig. 2.29 High-energy (>1 keV) electron trajectories with observed counts greater than 2 from one energy scan at 19:42:11–19:42:27 UT on 7 May 2009, traced back from the Kaguya position in a uniform magnetic field. The *black arrow* indicates the magnetic field direction obtained by LMAG

resulting magnetic charge distribution and the radial component at the Kaguya altitudes calculated from the magnetic charge distribution. These synthesized magnetic field data include lunar crustal fields with length scales greater than 20 km.

Figure 2.31 shows the electron trajectories obtained from tracing calculations that include the synthesized lunar magnetic field data. Compared with the uniform magnetic field case shown in Fig. 2.29, some of the electrons are traced back to outer space, suggesting that these electrons are scattered by the large-scale crustal fields. However, most of the electrons still originate from near the surface even if

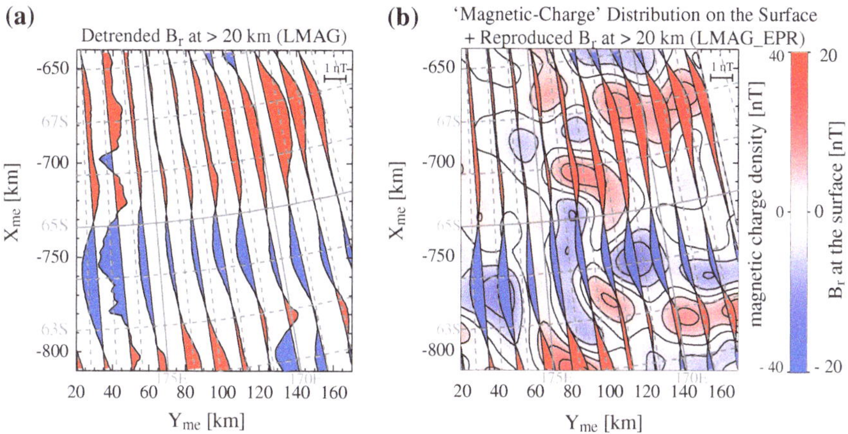

Fig. 2.30 Maps of the southwestern side of the South Pole–Aitken basin (orthographic projection). **a** The radial component of the detrended magnetic field data obtained by Kaguya MAP–LMAG at low altitudes in the terrestrial magnetotail. **b** The "magnetic charge" distribution on the surface (2 km spacing, not shown) derived by applying the EPR method to the detrended data, and the radial component at the Kaguya altitudes calculated from the magnetic charge distribution

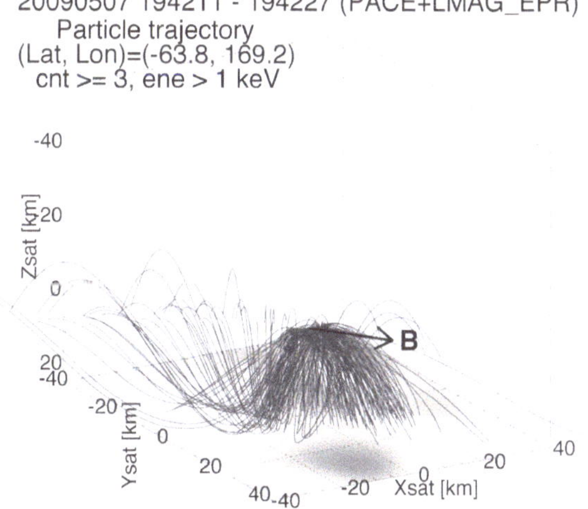

Fig. 2.31 High-energy (>1 keV) electron trajectories with observed counts greater than 2 from one energy scan at 19:42:11–19:42:27 UT on 7 May 2009, traced back from the Kaguya position in non-uniform magnetic fields including the synthesized lunar magnetic field data (cf. Fig. 2.30b). The *black arrow* indicates the magnetic field direction obtained by LMAG

large-scale magnetic fields are taken into account. This suggests a significant contribution to the electron trajectories from crustal fields that are absent from the synthesized large-scale magnetic field data; i.e., the small-scale magnetic fields (see Fig. 2.32).

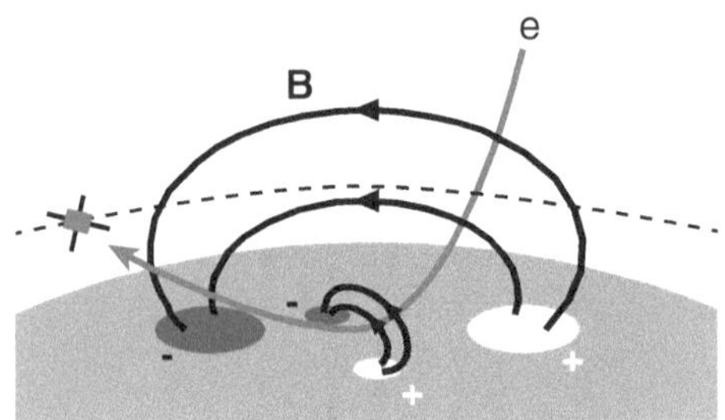

Fig. 2.32 Schematic illustration of high-energy electron trajectories non-adiabatically scattered by a small-scale crustal magnetic field; the *white* and *dark gray zones* on the surface indicate equivalent sources of outward and inward magnetic fields, respectively

It is possible to extract information about the small-scale magnetic fields on the lunar surface from these observations of high-energy electron scattering. If I assume that the scale length of the surface magnetic field is much smaller than the electron gyrodiameter (i.e., the non-adiabatic limit), the electron trajectory will be bent sharply as shown in Fig. 2.33. Since the magnetic force is always perpendicular to the electron velocity, a horizontal component of the surface magnetic field is necessary to modify the electron trajectory upward from the surface. For a given scattered velocity at the surface (cf. $\mathbf{v}_1$ in Fig. 2.33), a lower limit of electron momentum change caused by magnetic scattering can be derived by considering the electron trajectory where the deflection angle, $\Delta\psi$, is the smallest (cf. the dashed trajectory in Fig. 2.33). From the equation of motion of an electron with charge e and mass m_e in a magnetic field $\mathbf{B}$, $m_e d\mathbf{v}/dt = e\mathbf{v} \times \mathbf{B}$, the velocity change of the electron along this trajectory, Δv, satisfies the equation

$$m_e \frac{\Delta v}{L_{h\perp}/v} = evB_{h\perp}, \tag{2.11}$$

where $B_{h\perp}$ is the horizontal component of the surface magnetic field perpendicular to the electron velocity, $L_{h\perp}$ is its horizontal scale length, and v is the electron speed. The product $B_{h\perp}L_{h\perp}$ can be derived as

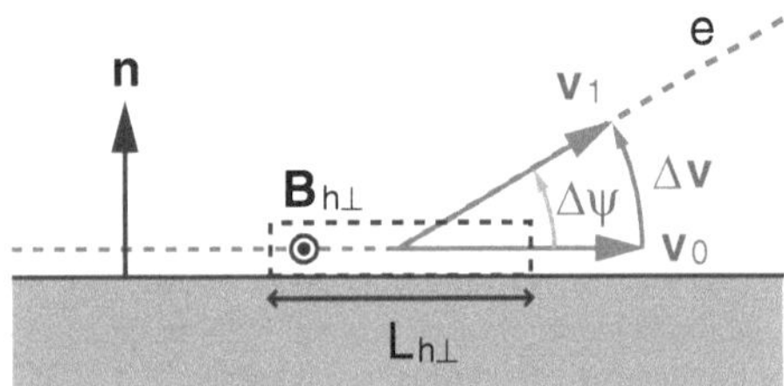

Fig. 2.33 Schematic illustration of high-energy electron trajectories passing through a small-scale horizontal magnetic field. $B_{h\perp}$ is the horizontal component of the surface magnetic field perpendicular to the electron velocity; $L_{h\perp}$ is its horizontal scale length; $\mathbf{n}$ is the normal direction to the lunar surface; $\mathbf{v}_0$ and $\mathbf{v}_1$ are the electron velocities before and after scattering, respectively; Δv is the velocity change; $\Delta\psi$ is the deflection angle; and the *dashed gray line* denotes the electron trajectory where Δv and $\Delta\psi$ are the smallest for a given $\mathbf{v}_1$

$$B_{h\perp} L_{h\perp} = \frac{m_e \Delta v}{e} = \frac{m_e v \Delta \psi}{e}. \tag{2.12}$$

A single electron-trace calculation gives a vector of $\mathbf{v}_1$, and I can easily calculate Δv and convert it to $B_{h\perp} L_{h\perp}$. Thus, a number of electron-tracing calculations can constrain the lower limit of the product of the horizontal magnetic field component and its horizontal scale length, $B_h L_h$, at a particular position on the lunar surface.

I now present the result of mapping small-scale horizontal magnetic fields in the southwestern side of the South Pole–Aitken (SPA) basin. I use electron data from Kaguya MAP–PACE during the same 12 orbits. I trace the electrons with energies >1 keV if the observed counts are greater than 2. The particle-tracing calculation includes both the lunar magnetic field data synthesized by the EPR method and the ambient magnetic field vector obtained by subtracting the synthesized magnetic field data from LMAG measurements at the time when the electron is observed. If the electron traced from the spacecraft strikes the lunar surface, its location and velocity at the surface are recorded. After analyzing all data from the 12 orbits, I divide the surface into 3 km bins and take the maximum value of Δv (or equivalently $B_{h\perp} L_{h\perp}$) in each bin. Since each surface velocity gives a lower limit for $B_h L_h$ at the point where the electron is scattered, I can constrain $B_h L_h$ for each bin by taking the maximum value. Figure 2.34a shows the mapping obtained with this procedure. The colors represent the lower limit of $B_h L_h$ that needs to be added to the synthesized large-scale field data to account for the electron observations.

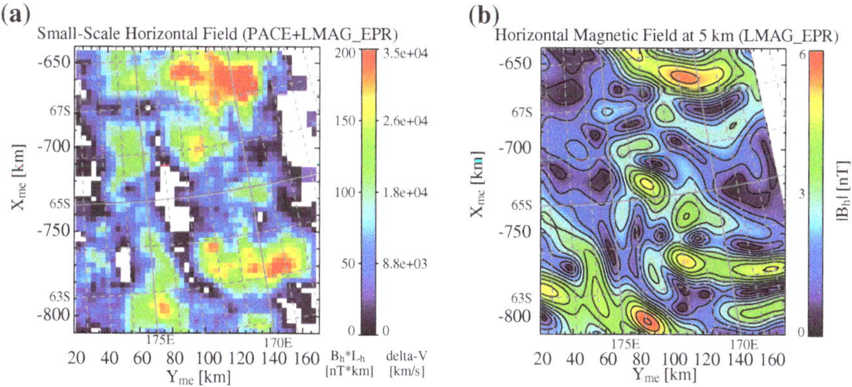

Fig. 2.34 Maps of the southwestern side of the South Pole–Aitken basin (orthographic projection). **a** Small-scale horizontal magnetic fields inferred from the combination of the electron data with energies >1 keV obtained from Kaguya MAP–PACE and the synthesized lunar magnetic field data (see text for details). The lower limit of the product of the horizontal magnetic field component and its horizontal scale length, $B_h L_h$, as well as that of the electron velocity change at the surface, Δv, is shown by *colors* in 3 km bins smoothed over 9 km (the bin size is chosen to achieve most complete coverage with good resolution). **b** Intensity of the horizontal component at 5 km altitude calculated from the magnetic charge distribution (cf. Fig. 2.30b)

2.4.2 Discussion

Here I briefly discuss the mapping of the small-scale magnetic fields on the lunar surface. The inferred $B_h L_h$ is up to 200 nT km. Since the synthesized lunar magnetic field data already include the large-scale magnetization (>20 km), I can reasonably assume $L_h < 20$ km. This yields the strongest $B_h > 10$ nT. Although this value is not surprisingly strong compared with the Apollo surface measurements of up to hundreds of nanotesla, our result suggests the existence of kilometer-scale magnetization with strength comparable to that of the large-scale magnetization (<20 nT as shown in Fig. 2.30b). While the Apollo and Lunokhod-2 surface measurements were all made on the Moon's nearside, I now find significant small-scale magnetization in the southwestern side of the SPA basin on the farside. This implies that kilometer-scale magnetization is not a unique characteristic of the nearside landing sites, but may be ubiquitous over the lunar surface.

Note that the small-scale horizontal magnetic fields inferred from the high-energy electron data seem to be correlated with the large-scale horizontal fields. Figure 2.34b shows the horizontal component of the large-scale lunar magnetic fields at 5 km altitude calculated from the magnetic charge distribution in Fig. 2.30b. The strong small-scale horizontal field regions indicated by the bright red colors in Fig. 2.34a tend to be found in regions where a strong large-scale horizontal magnetic field exists, as denoted by the warmer colors in Fig. 2.34b. This correlation suggests that the small-scale magnetization is not randomly distributed, but related to the large-scale magnetization. This result also implies that the observed keV electrons from the surface are crustal-field related, not caused by random transient features of the ambient field.

I note that high-energy electrons are expected to be less sensitive to near-surface electric fields than low-energy electrons. In fact, the negative surface potentials estimated from the upward electron-beam energies observed during this time period are a few hundred volts, insufficient to reflect the keV electrons. In addition to the well-established adiabatic reflection of low-energy electrons, non-adiabatic scattering of high-energy electrons can be used as a complementary method to investigate the small-scale component of the crustal magnetization.

2.5 Conclusions

Non-gyrotropic electron VDFs observed near the lunar surface (at altitudes of ~10–100 km) by Kaguya have been analyzed and compared with theoretical predictions from particle-trace calculations. Non-gyrotropic empty regions in the electron VDFs are produced by the gyro-loss effect, which is attributed to the absorption of electrons by the lunar surface, combined with electron gyromotion. Based on particle-trace calculations, I derived theoretical forbidden regions in the electron VDFs and considered the modifications to the forbidden regions caused by nonuniform magnetic fields associated with diamagnetic-current systems, lunar-surface charging, and electric fields oriented perpendicular to the magnetic field.

The non-gyrotropic electron VDF caused by the gyro-loss effect corresponds to a diamagnetic current seen in the electron VDF. The diamagnetic-current system is set up by the pressure gradient near the lunar surface and produces nonuniform magnetic-field configurations. By modeling non-gyrotropic electron VDFs for different altitudes, one-dimensional distributions of the magnetic-field intensity are derived, and the modifications to the forbidden regions are estimated. The diamagnetic-current effect on the forbidden regions depends on the ambient electron density, temperature, and magnetic-field strength.

Taking surface potentials into account in the particle-trace calculations, the lunar surface-charging effect on the non-gyrotropic forbidden regions is estimated. Negative surface potentials reflect the electrons and reduce the forbidden regions, depending on the electron energy. Also, calculations including the $\mathbf{E} \times \mathbf{B}$ drift term indicate that electric fields perpendicular to the magnetic fields modify the forbidden regions depending on the drift direction.

Comparison between the observed empty regions and the calculated forbidden regions suggests that various factors affect the characteristics of the non-gyrotropic electron VDFs, depending on the ambient-plasma conditions. In the magnetotail lobes with low density and temperature in relatively strong magnetic fields, the diamagnetic-current effect cannot work well. Perpendicular electric fields also seem to be less effective, because plasma flows are expected to be moderate in the tail lobes. Instead, on the lunar nightside, the lunar-surface potentials slightly modify the forbidden regions. On the dayside of the Moon in the solar wind, there will be no lunar surface-charging effect on the forbidden regions with positive surface potentials, and observations suggest a solar-wind motional electric-field effect and/or a diamagnetic-current effect in relatively weak magnetic-field conditions for high densities compared with those in the Earth's magnetosphere. In the terrestrial plasma sheet, all three mechanisms can modify the forbidden regions significantly. High electron temperature and weak magnetic fields result in a strong diamagnetic-current effect despite the relatively low density. Large surface potentials in the plasma sheet will reflect higher-energy electrons. Although convection electric fields do not seem to be able to modify electron trajectories for energies of a few keV, as analyzed in this chapter, they will affect lower-energy electrons that are lost at lower altitudes. One of the observations implies the presence of a local electric field of at least 5 mV/m, but the mechanism for producing this electric field is unknown The dominant modification processes deduced from the observed non-gyrotropic electron VDFs in different regions are summarized in Table 2.2.

Table 2.2 Summary of the dominant modification effects on non-gyrotropic electron VDFs observed in different regions

	Dayside	Nightside
Lobe	–	U_{M}
Solar wind	$E_{\perp}$ and/or $J_{\mathrm{D}e}$	–
Plasma sheet	–	$E_{\perp}$, $J_{\mathrm{D}e}$, and U_{M}

$E_{\perp}$, $J_{\mathrm{D}e}$, and U_{M} represent the perpendicular electric-field, diamagnetic-current, and lunar surface-charging effects, respectively

Some observations in the terrestrial plasma sheet show highly complex empty regions in electron VDFs, which can be explained by none of the above-mentioned mechanisms. A plausible explanation is that the plasma sheet electrons are non-adiabatically scattered by small-scale magnetic fields on the lunar surface. Combining these electron observations and particle tracing that includes the synthesized data of large-scale lunar magnetic fields, I developed a new mapping technique. The result implies significant kilometer-scale magnetic fields correlated with larger-scale magnetization on the Moon's far side.

The gyro-loss effect is a fundamental process associated with charged and magnetized particles and boundaries. This process will be distinctly observed, particularly around sharp boundaries such as provided by solid surfaces. Various components can affect the production of non-gyrotropic VDFs. In other words, analysis of non-gyrotropic VDFs might provide a wealth of information on plasma environments near surfaces. The basic process can be applied to other airless bodies such as Mercury, asteroids, other moons, and distant comets. High-angular-resolution observations of non-gyrotropic VDFs near solid surfaces can promote a better understanding of the near-surface electromagnetic environment and of plasma–solid-surface interactions.

References

1. Anderson, K.A., Lin, R.P., Gurgiolo, C., Parks, G.K., Potter, D.W., Werden, S., Rème, H.: A component of nongyrotropic (phase-bunched) electrons upstream from the Earth's bow shock. J. Geophys. Res. **90**(NA11), 10,809–10,814 (1985). doi:10.1029/JA090iA11p10809
2. Anderson, K.A., Lin, R.P., McGuire, R.E., McCoy, J.E.: Measurement of lunar and planetary magnetic fields by reflection of low energy electrons. Space Sci. Instrum. **1**, 439–470 (1975)
3. Armstrong, T.P., Choo, T.H., Roelof, E.C., Simnett, G.M., Sarris, E.T.: The effect of the shock of 15:43 UT March 23, 1991 on 50 keV to 5 MeV ions at Ulysses. Geophys. Res. Lett. **19**(12), 1247–1250 (1992). doi:10.1029/92GL00777
4. Coates, A.J., Jones, G.H.: Plasma environment of Jupiter family comets. Planet. Space Sci. **57**(10, Sp. Iss. SI), 1175–1191 (2009). doi:10.1016/j.pss.2009.04.009
5. Coleman, P.J., Russell, C.T., Sharp, L.R., Schubert, G.: Preliminary mapping of the lunar magnetic field. Phys. Earth Planet. Inter. **6**, 167–174 (1972). doi:10.1016/0031-9201(72)90050-7
6. Collinson, D.W.: Magnetism of the Moon—a lunar core dynamo or impact magnetization? Surv. Geophys. **14**(1), 89–118 (1993). doi:10.1007/BF01044078
7. Dolginov, S., Yeroshenko, Y., Sharova (Izmiran), V., Vnuchkova (Ikian), T., Vanyan, L., Okulessky (Ioan), B., Bazilevsky (Geochian), A.: Study of magnetic field, rock magnetization and lunar electrical conductivity in the bay le monnier. Moon **15**, 3–14 (1976). doi:10.1007/BF00562468. http://dx.doi.org/10.1007/BF00562468
8. Dyal, P., Parkin, C.W., Daily, W.D.: Magnetism and the interior of the Moon. Rev. Geophys. Space Phys. 12, 568–591 (1974). doi:10.1029/RG012i004p00568
9. Dyal, P., Parkin, C.W., Snyder, C.W., Clay, D.R.: Measurements of lunar magnetic field interaction with the solar wind. Nature **236**(5347), 381–385 (1972). doi:10.1038/236381a0
10. Farrell, W.M., Stubbs, T.J., Vondrak, R.R., Delory, G.T., Halekas, J.S.: Complex electric fields near the lunar terminator: the near-surface wake and accelerated dust. Geophys. Res. Lett. **34**, L14,201 (2007). doi:10.1029/2007GL029312

11. Futaana, Y., Machida, S., Saito, Y., Matsuoka, A., Hayakawa, H.: Moon-related nonthermal ions observed by Nozomi: species, sources, and generation mechanisms. J. Geophys. Res. **108**, 1025 (2003). doi:10.1029/2002JA009366
12. Gurgiolo, C., Larson, D., Lin, R.P., Wong, H.K.: A gyrophase-bunched electron signature upstream of the Earth's bow shock. Geophys. Res. Lett. **27**, 3153–3156 (2000). doi:10.1029/2000GL000065
13. Halekas, J.S., Delory, G.T., Lin, R.P., Stubbs, T.J., Farrell, W.M.: Lunar prospector observations of the electrostatic potential of the lunar surface and its response to incident currents. J. Geophys. Res. **113**, A09,102 (2008). doi:10.1029/2008JA013194
14. Halekas, J.S., Lillis, R.J., Lin, R.P., Manga, M., Purucker, M.E., Carley, R.A.: How strong are lunar crustal magnetic fields at the surface? Considerations from a reexamination of the electron reflectometry technique. J. Geophys. Res. **115**, E03,006 (2010). doi:10.1029/2009JE003516
15. Halekas, J.S., Mitchell, D.L., Lin, R.P., Frey, S., Hood, L.L., Acuña, M.H., Binder, A.B.: Mapping of crustal magnetic anomalies on the lunar near side by the Lunar Prospector electron reflectometer. J. Geophys. Res. **106**, 27,841–27,852 (2001). doi:10.1029/2000JE001380
16. Hood, L.L., Zakharian, A., Halekas, J.S., Mitchell, D.L., Lin, R.P., Acuña, M.H., Binder, A.B.: Initial mapping and interpretation of lunar crustal magnetic anomalies using Lunar Prospector magnetometer data. J. Geophys. Res. **106**, 27,825–27,839 (2001). doi:10.1029/2000JE001366
17. Kivelson, M.G., Russell, C.T.: Introduction to Space Physics. Cambridge University Press, Cambridge (1995)
18. Kurata, M., Tsunakawa, H., Saito, Y., Shibuya, H., Matsushima, M., Shimizu, H.: Mini-magnetosphere over the Reiner Gamma magnetic anomaly region on the Moon. Geophys. Res. Lett. **32**(24), L24,205 (2005). doi:10.1029/2005GL024097
19. Leubner, M.: An analytical representation of non-gyrotropic distributions and related space applications. Planet. Space Sci. **51**(12), 723–729 (2003). doi:10.1016/S0032-0633(03)00109-0
20. Lin, R.: Constraints on the origins of lunar magnetism from electron reflection measurements of surface magnetic fields. Phys. Earth Planet. Inter. **20**(2–4), 271–280 (1979). doi:10.1016/0031-9201(79)90050-5
21. Lin, R.P., Anderson, K.A., Hood, L.L.: Lunar surface magnetic field concentrations antipodal to young large impact basins. Icarus **74**, 529–541 (1988). doi:10.1016/0019-1035(88)90119-4
22. Lin, R.P., Mitchell, D.L., Curtis, D.W., Anderson, K.A., Carlson, C.W., McFadden, J., Acuña, M.H., Hood, L.L., Binder, A.: Lunar surface magnetic fields and their interaction with the solar wind: results from Lunar Prospector. Science **281**, 1480–1484 (1998). doi:10.1126/science.281.5382.1480
23. Lue, C., Futaana, Y., Barabash, S., Wieser, M., Holmström, M., Bhardwaj, A., Dhanya, M.B., Wurz, P.: Strong influence of lunar crustal fields on the solar wind flow. Geophys. Res. Lett. **38**, L03,202 (2011). doi:10.1029/2010GL046215
24. Matsushima, M., Tsunakawa, H., Iijima, Y.i., Nakazawa, S., Matsuoka, A., Ikegami, S., Ishikawa, T., Shibuya, H., Shimizu, H., Takahashi, F.: Magnetic cleanliness program under control of electromagnetic compatibility for the SELENE (Kaguya) spacecraft. Space Sci. Rev. **154**(1–4), 253–264 (2010). doi:10.1007/s11214-010-9655-x
25. McGuire, R.E.: A theoretical treatment of lunar particle shadows. Cosmic Electrodynamics **3**, 208–239 (1972)
26. Meziane, K., Lin, R., Parks, G., Larson, D., Bale, S., Mason, G., Dwyer, J., Lepping, R.: Evidence for acceleration of ions to similar to 1 MeV by adiabatic-like reflection at the quasi-perpendicular Earth's bow shock. Geophys. Res. Lett. **26**(19), 2925–2928 (1999). doi:10.1029/1999GL900603
27. Mitchell, D.L., Halekas, J.S., Lin, R.P., Frey, S., Hood, L.L., Acuña, M.H., Binder, A.: Global mapping of lunar crustal magnetic fields by Lunar Prospector. Icarus **194**, 401–409 (2008). doi:10.1016/j.icarus.2007.10.027
28. Mukai, T., Yamamoto, T., Machida, S.: Dynamics and kinetic properties of plasmoids and flux ropes: GEOTAIL observations. Geophys. Monogr. Ser. **105**, 117–137 (1998). doi:10.1029/GM105p0117

29. Purucker, M.E.: A global model of the internal magnetic field of the Moon based on Lunar Prospector magnetometer observations. Icarus **197**(1), 19–23 (2008). doi:10.1016/j.icarus.2008.03.016
30. Purucker, M.E., Iii, J.W.H., Wilson, L.: Magnetic signature of the lunar south pole-aitken basin: character, origin, and age. J. Geophys. Res. **117**, E05,001 (2012). doi:10.1029/2011JE003922
31. Purucker, M.E., Nicholas, J.B.: Global spherical harmonic models of the internal magnetic field of the Moon based on sequential and coestimation approaches. J. Geophys. Res. **115**, E12,007 (2010). doi:10.1029/2010JE003650
32. Reiff, P.H.: Magnetic shadowing of charged particles by an extended surface. J. Geophys. Res. **81**(19), 3423–3427 (1976). doi:10.1029/JA081i019p03423
33. Reiff, P.H., Burke, W.J.: Interactions of the plasma sheet with the lunar surface at the Apollo 14 site. J. Geophys. Res. **81**(25), 4761–4764 (1976). doi:10.1029/JA081i025p04761
34. Saito, Y., Yokota, S., Asamura, K., Tanaka, T., Akiba, R., Fujimoto, M., Hasegawa, H., Hayakawa, H., Hirahara, M., Hoshino, M., Machida, S., Mukai, T., Nagai, T., Nagatsuma, T., Nakamura, M., Ichiro Oyama, K., Sagawa, E., Sasaki, S., Seki, K., Terasawa, T.: Low-energy charged particle measurement by MAP-PACE onboard SELENE. Earth Planets Space **60**, 375–385 (2008)
35. Saito, Y., Yokota, S., Asamura, K., Tanaka, T., Nishino, M.N., Yamamoto, T., Terakawa, Y., Fujimoto, M., Hasegawa, H., Hayakawa, H., Hirahara, M., Hoshino, M., Machida, S., Mukai, T., Nagai, T., Nagatsuma, T., Nakagawa, T., Nakamura, M., Ichiro Oyama, K., Sagawa, E., Sasaki, S., Seki, K., Shinohara, I., Terasawa, T., Tsunakawa, H., Shibuya, H., Matsushima, M., Shimizu, H., Takahashi, F.: In-flight performance and initial results of plasma energy angle and composition experiment (PACE) on SELENE (Kaguya). Space Sci. Rev. **154**, 265–303 (2010). doi:10.1007/s11214-010-9647-x
36. Shimizu, H., Takahashi, F., Horii, N., Matsuoka, A., Matsushima, M., Shibuya, H., Tsunakawa, H.: Ground calibration of the high-sensitivity SELENE Lunar Magnetometer LMAG. Earth Planets Space **60**, 353–363 (2008)
37. Takahashi, F., Shimizu, H., Matsushima, M., Shibuya, H., Matsuoka, A., Nakazawa, S., Iijima, Y., Otake, H., Tsunakawa, H.: In-orbit calibration of the Lunar Magnetometer onboard SELENE (KAGUYA). Earth Planets Space **61**, 1269–1274 (2009)
38. Toyoshima, M., Shibuya, H., Matsushima, M., Shimizu, H., Tsunakawa, H.: Equivalent source mapping of the lunar crustal magnetic field using ABIC. Earth Planets Space **60**(4), 365–373 (2008). First SELENE Working Team Meeting, Tsukuba Space Ctr, Tsukuba, JAPAN, 09–11 Jan 2007
39. Tsunakawa, H., Shibuya, H., Takahashi, F., Shimizu, H., Matsushima, M., Matsuoka, A., Nakazawa, S., Otake, H., Iijima, Y.: Lunar magnetic field observation and initial global mapping of lunar magnetic anomalies by MAP-LMAG onboard SELENE (Kaguya). Space Sci. Rev. **154**, 219–251 (2010). doi:10.1007/s11214-010-9652-0
40. Tu, J., Mukai, T., Hoshino, M., Saito, Y., Matsuno, Y., Yamamoto, T., Kokubun, S.: Geotail observations of ion velocity distributions with multi-beam structures in the post-plasmoid current sheet. Geophys. Res. Lett. **24**(17), 2247–2250 (1997). doi:10.1029/97GL02113
41. Wieser, M., Barabash, S., Futaana, Y., Holmström, M., Bhardwaj, A., Sridharan, R., Dhanya, M.B., Schaufelberger, A., Wurz, P., Asamura, K.: First observation of a mini-magnetosphere above a lunar magnetic anomaly using energetic neutral atoms. Geophys. Res. Lett. **37**, L05,103 (2010). doi:10.1029/2009GL041721

Chapter 3
Lunar Dayside Plasma in the Earth's Magnetotail Lobes

Abstract As introduced in Chap. 1, the lunar atmosphere is too tenuous to stand off the solar wind plasma and magnetic field by inducing currents in its ionosphere. Therefore, the Moon is often called an "airless" body and thought of as a passive plasma absorber. This chapter presents the dual-probe ARTEMIS observations in the Earth's magnetotail lobes to demonstrate that plasma of lunar origin can be dominant and have a significant impact on the ambient plasma. The two-point measurements reveal that the plasma density on the lunar dayside sometimes becomes several times higher than the ambient lobe plasma density. Meanwhile, the electron pitch-angle distributions show $\sim$90° electron dropouts coexisting with the plasma of lunar origin, suggesting that the velocity distributions of lobe electrons are modified in association with the lunar plasma. The accurate electron measurements by ARTEMIS with knowledge of the spacecraft potential also suggest the existence of a high-energy tail population of lunar surface photoelectrons. The high-energy photoelectron emission results in large positive potentials on the dayside lunar surface in the tail lobes, accelerating lunar ions upward from the Moon.

Keywords Plasma of lunar origin · Terrestrial magnetotail lobes · Lunar surface photoelectrons

3.1 Introduction

In this chapter, I describe Moon-related electron and ion signatures observed by ARTEMIS in the terrestrial magnetotail lobe. Chapter 2 deals with one-point measurements at low altitudes (<100 km), while this chapter presents dual-probe measurements that cover a wide spatial range from a few hundred km up to $\sim$20,000 km distant from the lunar surface. Even though the spacecraft altitudes are much larger than the electron gyroradius, ARTEMIS can still observe the gyro-loss effect on upcoming electrons from the lunar surface. In addition, analysis of the ARTEMIS data helps resolve two unsolved issues regarding the lunar plasma environment:

Y. Harada, *Interactions of Earth's Magnetotail Plasma with the Surface, Plasma, and Magnetic Anomalies of the Moon*, Springer Theses,
DOI 10.1007/978-4-431-55084-6_3

- Do high-energy tail populations of lunar surface photoelectrons exist?
- Can plasma of lunar origin have significant effects on the near-lunar environment?

The ARTEMIS mission [3] conducts comprehensive plasma and wave observations by employing two probes that orbit the Moon. Using simultaneous observations by probes that are, respectively, near and distant from the Moon, Moon-related phenomena can be separated from those intrinsic to the ambient plasma. The ARTEMIS measurements of electron velocity distributions, utilizing knowledge of the spacecraft potential, enables us for the first time to quantitatively interpret the energy spectra of upward-traveling electrons that play the crucial role in the surface charging mechanisms. I first show the pitch-angle and energy distributions of field-aligned, upward-traveling electrons from the dayside lunar surface and discuss their generation processes and implications for large positive potentials of the lunar surface. I then present dual-probe data on Moon-related populations observed above the lunar dayside, which confirm the large positive surface potentials and indicate that plasma of lunar origin dominates the lobe plasma populations that are also present. In addition, I present modified electron-velocity distribution functions associated with Moon-related populations. These ARTEMIS observations reveal an interesting plasma environment, where dominant heavy ions are accelerated near the positively charged surface, expanding several hundreds of kilometers above the Moon's dayside and potentially interacting with the ambient-plasma populations.

3.2 Field-Aligned Electrons from the Dayside Lunar Surface

I first present the ARTEMIS data of the field-aligned, upward-traveling electrons from the dayside lunar surface observed in the terrestrial magnetotail lobe. Analysis of the pitch-angle and energy distributions of the upward-traveling electrons provides important pieces of information on the near-surface dynamics of electrons, as well as on the electrostatic potentials of the dayside lunar surface. The lunar surface potentials, which are mainly determined by electron dynamics, have significant implications for the near-surface dynamics of newborn ions of lunar exospheric origin.

Figure 3.1 shows the orbits of the two ARTEMIS probes and the average magnetic-field direction during the ARTEMIS P2 lunar dayside observations in the terrestrial magnetotail on 7 January 2012. Figure 3.2 shows the data from the ARTEMIS P2 during this time interval, based on data from the ESA (ElectroStatic Analyzer) [15], EFI (Electric Field Instrument) [5], and FGM (Fluxgate Magnetometer) [4] instruments. Hot electrons and ions with energies greater than 1 keV were not observed, and the magnetic-field data show a coherent, dominant, and positive B_x component, indicating that the Moon was located in the northern tail lobe. Straight-line magnetic-field traces from the spacecraft imply a magnetic connection between P2 and the dayside lunar surface at 17:55–18:25. As seen in the electron pitch-angle distributions, a considerable upward flux of electrons with pitch angles between 0° and 90° was detected during this magnetically connected interval. The upward-traveling

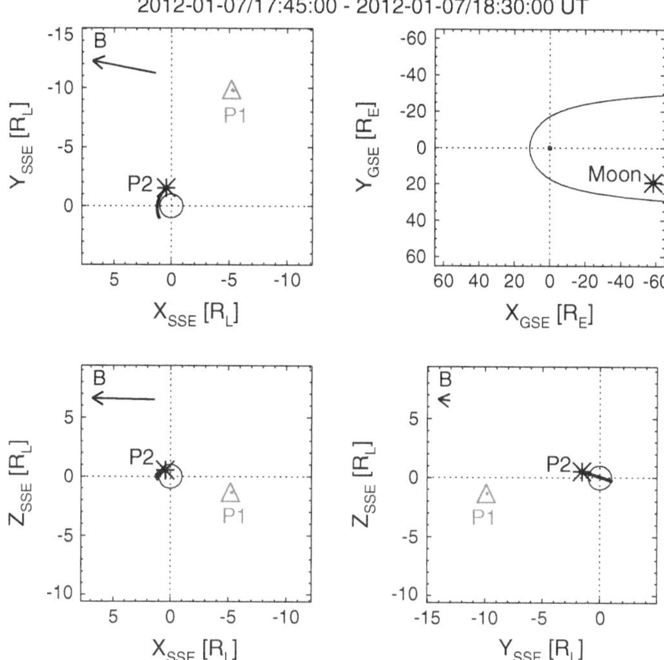

Fig. 3.1 ARTEMIS P1 (*gray*) and P2 (*black*) orbits for the event shown in Figs. 3.2 and 3.5, with the initial locations indicated by the *gray triangles* and the *black asterisks*, respectively (the lunar radius $R_L = 1{,}738$ km and the Earth radius $R_E = 6{,}378$ km). The black asterisk in the top right-hand panel shows the Moon's location. The selenocentric solar ecliptic (SSE) system has its X-axis from the Moon toward the Sun, the Z-axis is parallel to the upward normal to the Earth's ecliptic plane, and and the Y completes the orthogonal coordinate set, whereas the geocentric solar ecliptic (GSE) system has its X-axis from the Earth toward the Sun. The black arrows show the magnetic-field directions averaged over the orbits. The black thin line in the top right-hand panel shows the typical location of the magnetopause in GSE coordinates, indicating that the Moon was located in the magnetotail

electrons exhibit a field-aligned signature, particularly at the beginning and end of the connected interval.

The field-aligned signature of the upward-traveling electrons from the dayside lunar surface can be explained by considering electron emission from an extended surface in an oblique magnetic field, as sketched in Fig. 3.3a. Electrons with pitch angles (α) smaller than the elevation angle (δ) of the magnetic field from the surface will be emitted with all 360° gyrophases, as shown by the dark gray, complete cone. They will all travel upward along the field line without striking the surface, because their trajectories are spatially restricted within a cone of angle δ from the field line (cf. the black trajectory in Fig. 3.3b). On the other hand, the initial gyrophases of electrons with $\delta < \alpha < 90°$ are limited because of the presence of the surface, as shown by the light gray, partial cone in Fig. 3.3a. Some will strike the surface

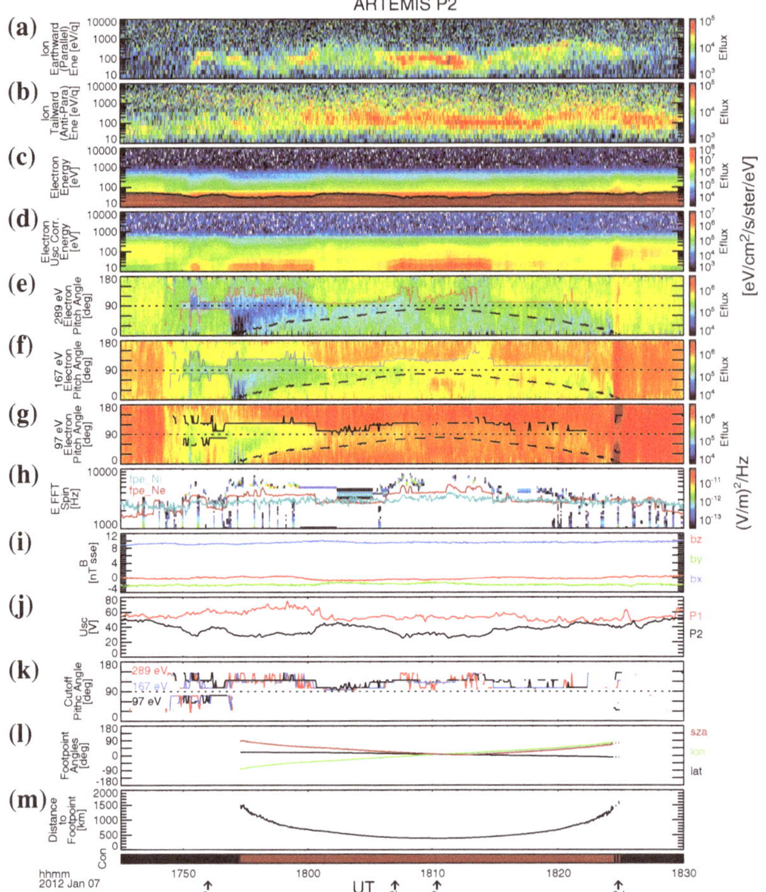

Fig. 3.2 Time-series data from an ARTEMIS P2 lunar-dayside flyby in the magnetotail lobe on 7 January 2012. Differential energy-flux spectra for ions with pitch angles of **a** 0–90° (Earthward, corresponding to the upward motion from the Moon's dayside) and **b** 90–180° (tailward, corresponding to the downward motion toward the Moon's dayside), for omnidirectional electrons **c** with the spacecraft potential indicated by the black line and **d** with the spacecraft potential corrected, **e**–**g** electron pitch-angle spectra for three energy channels, **h** high-frequency electric-field spectra from the spin-plane sensors, **i** magnetic field in SSE coordinates, **j** spacecraft potentials from P2 and P1, **k** cutoff pitch angles derived from the electron pitch-angle distribution averaged over three spin periods, **l** solar zenith angle, longitude, and latitude of the foot point, and **m** distance along the field line to the foot point. The *dashed lines* in panels (**e**)–(**g**) indicate the elevation angle δ of the magnetic field from the lunar surface estimated from straight-line magnetic-field extrapolations. The *red*, *blue*, and *black lines* in panels (**e**), (**f**), and (**g**) show the cutoff pitch angle at each energy corresponding to the lines in panel (**k**). The *light-blue* and *red lines* in panel (**h**) show the electron plasma frequencies calculated from the moments of the observed ion and electron distributions, respectively. The *bottom color bar* indicates magnetic connection to the lunar surface in *red* and no connection in black. The *four arrows* below the time axis denote the timing at which the electron distributions shown in Fig. 3.8 were obtained

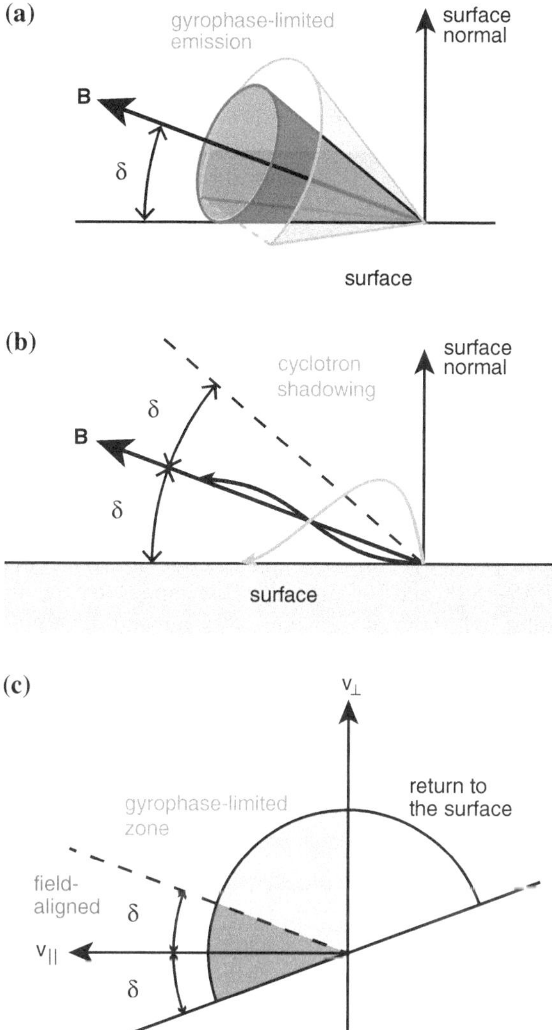

Fig. 3.3 Schematic illustration of **a** electron emission from an extended surface in an oblique magnetic field, **b** cyclotron shadowing by the surface, and **c** the velocity distribution function of the emitted electrons

because of gyromotion around the oblique field line (cf. the light gray trajectory in Fig. 3.3b). This shadowing effect of magnetized particles by an absorbing surface, which is called "cyclotron shadowing" [21] or the "gyro-loss" effect described in Chap. 2, depends on the initial gyrophase of the emitted electrons. This cyclotron shadowing further reduces the range of the emitted gyrophases with which electrons can travel upward to outer space along the field line. Electrons emitted with

$\alpha > 90°$ are unable to travel upward along the field line, and they will all return to the surface. Eventually, a boundary will appear at δ in the upward-traveling pitch-angle distribution, where the oblique magnetic field starts to be effective in restricting the initial gyrophases (see Fig. 3.3c). For example, even if the electron emission from the surface is isotropic, a lunar orbiter will detect the gyrophase-averaged electron flux which is reduced at $\delta < \alpha < 90°$ compared with at $0° < \alpha < \delta$ because of both gyrophase-limited emission and cyclotron shadowing, resulting in a field-aligned signature of the upward-traveling electrons.

I now compare the electron data with the prediction of the oblique magnetic-field effect. The magnetic-elevation angle δ can be estimated from the straight-line magnetic-field traces. First, the measured magnetic-field vector is extrapolated from the spacecraft until the field line crosses the lunar surface. I then compute δ based on the geometry of the magnetic-field vector relative to the surface normal at the magnetic footpoint. The dashed lines overplotted on the electron pitch-angle spectra in Fig. 3.2e–g indicate the estimated elevation angles δ. At each energy, a relatively large flux is observed for $0° < \alpha < \delta$, whereas the flux decreases for $\delta < \alpha < 90°$. The observed pitch-angle distributions of the upward electrons are in good agreement with the theoretical prediction of the gyrophase-limited zone for $\delta < \alpha < 90°$, particularly at 17:55–18:00 and 18:20–18:25. This implies that the oblique magnetic field gives rise to the field-aligned signature of the upward-traveling electrons.

The agreement of the observed pitch-angle spectra with the theoretical prediction is rather indistinct at 18:00–18:20. There are several possible reasons why it is difficult to observe the oblique magnetic-field effect during this time period. First, if upward electrons include magnetically-reflected populations (i.e., "loss cone" distributions), the reflected electrons with large flux will "mask" the smaller-flux electrons emitted from the lunar surface. The spot-like bright signatures seen in the upward-electron distributions at 18:02, 18:10–18:13, 18:15, 18:17, and 18:18 (see Fig. 3.2e–g) are presumably due to electron reflection by lunar crustal magnetic fields. These features will be discussed in more detail in Sect. 3.4. Second, electron emission from the lunar surface might be anisotropic. As δ approaches 90°, the oblique-field effect becomes less effective. If the emitted-electron distribution is originally field-aligned, for example, it will be difficult to identify the boundary at δ in the pitch-angle distribution.

Another interesting issue regarding the field-aligned electrons from the dayside lunar surface in the tail lobes relates to their energy spectra. Figure 3.4a shows the energy distributions of the field-aligned, upward- and downward-traveling electrons in the energy range < 200 eV obtained by P2 at 18:05–18:06, whereas Fig. 3.4b shows those corrected for the spacecraft potential. One can see that large-flux spacecraft/instrument photoelectrons are successfully removed from the electron energy spectra.

Here I compare the field-aligned fluxes of upward- and downward-electrons. If I assume that the electron distribution over the entire velocity space is obtained and ion currents are negligible, the sum of the upward- and downward-electron currents should balance in the equilibrium state. The upward- and downward-flux along the field line, F_u and F_d, can be calculated as $F_u = \iint J|\cos\alpha| d\Omega dE$ for $0° < \alpha < 90°$,

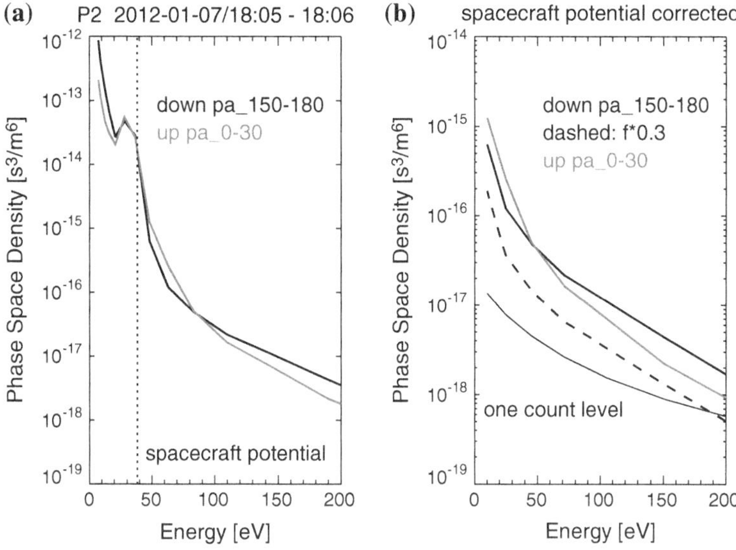

Fig. 3.4 Electron-energy spectra in units of distribution function obtained by P2 at 18:05–18:06 UT on 7 January 2012. The *black* and *gray thick lines* show electrons with pitch angles of 150–180° (*downward*) and 0–30° (*upward*), respectively. **a** The obtained energy spectra include contamination from spacecraft/instrument photoelectrons with energies lower than the spacecraft potential indicated by the *vertical dotted line*. **b** The spacecraft potential has been subtracted. It is seen that the spacecraft/instrument photoelectrons are successfully removed from the energy spectra. The dashed line shows 30 % of the downward-traveling electron spectrum. The upward distribution of backscattered electrons is expected to be lower than this line. The *black thin line* indicates the one-count level

and $F_d = \iint J|\cos\alpha| d\Omega dE$ for $90° < \alpha < 180°$, where J is the differential flux, E the energy, and Ω the solid angle. I use the electron-energy spectrum from P2 at 18:05–18:06 corrected for the spacecraft potential in units of distribution function, and convert it to differential flux. Integrating over the available energy range and each hemisphere, I obtain $F_u = 6.8 \times 10^6$ cm^{-2}s^{-1} and $F_d = 7.5 \times 10^6$ cm^{-2}s^{-1}. The integrated upward-electron flux is slightly smaller than the integrated downward-electron flux. The discrepancy can be attributed to the lowest-energy population just above the spacecraft potential that ESA cannot resolve and/or contribution from ion currents. For example, lack of measurements of upward-traveling cold electrons with very low energies will result in underestimation of F_u.

In Fig. 3.4b, the upward-traveling distribution (cf. the gray line) is much higher than the one-count level (cf. the thin black line). The fact that the upward-traveling electrons still have such a large population, even with relatively high energies of ~10–200 eV, can constrain their source mechanisms. As summarized in Table 1.2, the dayside lunar surface emits electrons mainly via three types of processes: photoemission, secondary electron emission, and backscattering. Impinging electrons that are reflected by magnetic or electric fields above the lunar surface may also exist.

Generally, secondary electrons and core populations of photoelectrons are expected to be very cold, with a characteristic temperature of only a few eV [28]. These cold populations are incompatible with the relatively hot energy distribution of upward-traveling electrons with energies of ~10–200 eV. Backscattering electrons, on the other hand, can be emitted with energies comparable to the incident electron energies. However, the backscattering ratio is expected to be less than 0.3 for ~100 eV incident electrons [28]. This fraction is insufficient to explain the observed upward population significantly larger than 30 % of the downward electrons (indicated by the dashed line in Fig. 3.4b).

Possible explanations for the upward electrons with energies of ~10–200 eV include (i) high-energy tail populations of lunar surface photoelectrons, (ii) incident electrons electrically reflected by a nonmonotonic potential structure near the dayside lunar surface [7, 16–18], and (iii) incident electrons adiabatically or non-adiabatically reflected by the lunar crustal magnetic fields [2, 8]. I note that the upward electrons with energies <50 eV have a larger flux than the downward electrons (see Fig. 3.4b). This excess flux cannot be explained by any reflection processes which conserve particle energies, including electrostatic and magnetic reflection. Therefore, I suspect that at least the lower-energy population of the upward electrons with energies of ~10–50 eV mainly consists of high-energy photoelectrons after deceleration through a potential just above the dayside lunar surface.

The presence of a large population of upward electrons with relatively high energies of ~10–200 eV has significant implications for the current balance on the lunar surface. In the tail lobes, the electron current into the lunar surface from the ambient plasma is very small. Therefore, most upward electrons should return to the surface to maintain the current balance at the surface. The existence of a large population of upward-traveling high-energy electrons means that large positive potentials are required to attract them back to the surface within the photoelectron sheath just above the surface and reach an equilibrium.

The large positive potentials of the lunar surface will play an important role in the dynamics of newly ionized ions of lunar exospheric origin near the dayside lunar surface. These ions can interact with the lunar surface sheath in two ways: (1) ions ionized within the sheath will gain acceleration from the surface field until they exit the sheath (where magnetotail convection then solely controls the ion motion), and (2) ions ionized near the Moon yet outside the lunar photoelectron sheath that are convected into the Moon by the magnetotail convection electric field can be repelled by the surface field and "scattered" along the magnetotail field lines. The strength of the positive surface potential directly correlates to the energy that these ions can gain, and in turn, determines how far above the lunar surface the ions can travel before they are convected away by the geomagnetic field [19]. In fact, Kaguya (at an altitude of 100 km) and ARTEMIS (at several hundred kilometers above the Moon) have detected Moon-related ions which were presumably accelerated by the upward electric field caused by large positive potentials of the dayside lunar surface in the tail lobes [20, 26]. I present a pickup-ion observation which supports the suggestion of large positive surface potentials in the next section.

3.3 Moon-Related Ions and Electrons on the Lunar Dayside in the Tail Lobes

In this section, I compare the data from the dual-probe measurement on the lunar dayside and of ambient plasma, which indicate the existence of dominant ions and electrons of lunar origin with respect to the ambient-plasma populations in the tail lobes. The top two panels in Figs. 3.2 and 3.5 show the Earthward- (parallel to the magnetic field in the northern lobe) and tailward-traveling (anti-parallel) ions. These ion spectra from P2 and P1 are also shown in Fig. 3.6. P1 was located ~10 lunar radii from the Moon (see Fig. 3.1) and detected tailward ions with energies of ~100 eV, indicating a cold ion flow from the Earth in the tail lobes (see Fig. 3.6e). Earthward-traveling ions were not detected by P1 (see Fig. 3.6d). On the other hand, P2 detected both Earthward- and tailward-traveling ions above the dayside lunar surface (see Fig. 3.6a–b). These Earthward ions detected by P2 travel upward from the Moon, suggesting a Moon-related origin.

I now present the ion velocity distributions in more detail. The ion pitch-angle distributions from P2 and P1 are shown in Fig. 3.6c, f. The Moon-related ions detected by P2 have different velocity components compared with the ambient ion flow detected by P1. The Moon-related ions have pitch angles between 0° and 90°, but not necessarily either 0° or 90°. These ions with intermediate pitch angles have both parallel and perpendicular velocity components [20], implying acceleration by parallel and perpendicular electric fields, such as by the lunar photoelectron sheath field (both parallel and perpendicular) and the magnetospheric convection electric field (only perpendicular). Considering the observed pickup-ion pitch angles of ~60° and energies of ~100 eV, and a spacecraft potential of ~+30 V at 18:11, one can place a lower limit to the positive lunar surface potential of ~+32.5 V by assuming that the near-surface electric field is oriented parallel to the magnetic field. I discussed the implications for large positive surface potentials in relation to the energy spectrum of the upward-traveling electrons in Sect. 3.2. They are consistent with the nonzero parallel-velocity component of the pickup ions deduced here. I note that it is possible to constrain the ion species from the energy and pitch-angle spectra of pickup ions by assuming a reasonable convection speed or charge neutrality, even though ARTEMIS probes are not capable of mass resolution [20, 29].

Observations of the electron-plasma frequency, f_{pe}, can be used as a useful diagnostic tool for the local electron density, and also for the ion density on the assumption of charge neutrality. The electric-field wave spectra from P2 show that the frequencies of the observed narrow-band plasma oscillations, also known as Langmuir waves, occasionally rise up to well above the f_{pe} deduced from the moments of the observed ion distributions (see Fig. 3.2h). This implies the existence of undetected low-energy ions that are repelled by the large positive spacecraft potential in the tail lobes, and/or of heavy ions. Since the moment calculation is based on the "proton-only" assumption, the density fraction of heavy ions will be underestimated by a factor $\sqrt{M}$, where M is the ion-to-proton mass ratio [14]. The observed frequencies are also higher than the f_{pe} estimated from the moments of the measured electron distributions. This is

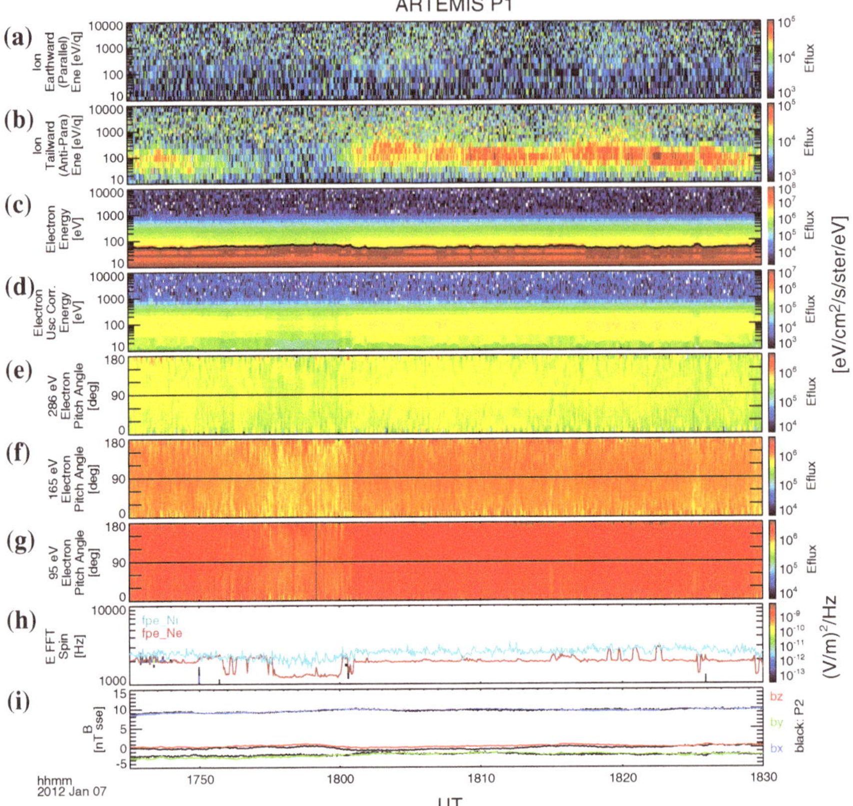

Fig. 3.5 Time-series data from ARTEMIS P1 during the same time interval as shown in Figs. 3.2 and 3.1. All panels have the same format as Fig. 3.2a–i, except that the black lines in panel (*i*) show the magnetic-field data from ARTEMIS P2

probably because the energy resolution around the spacecraft potential is not sufficiently high to resolve the enhanced cold electrons on the lunar dayside, nor to derive an accurate density from the moment calculation. The detected frequencies go up to the highest detectable frequency of 8 kHz at 18:08, indicating a corresponding plasma density of $\sim$0.8 cm^{-3}. On the other hand, the electron-plasma frequencies detected and estimated by P1 remain <4 kHz during the same time interval (see Fig. 3.5h). Thus, the electric field and particle data from P1 indicate ambient-plasma densities of less than $\sim$0.2 cm^{-3}. The discrepancy between the plasma densities from P2 and P1 implies the presence of several times as many ions and electrons above the lunar dayside as in the ambient plasma.

The measured spacecraft potential provides another proxy for the plasma density. P2 detected a spacecraft potential decrease that was correlated with the electron-plasma frequency's increase, whereas the P1 potential remained high (see Fig. 3.2j).

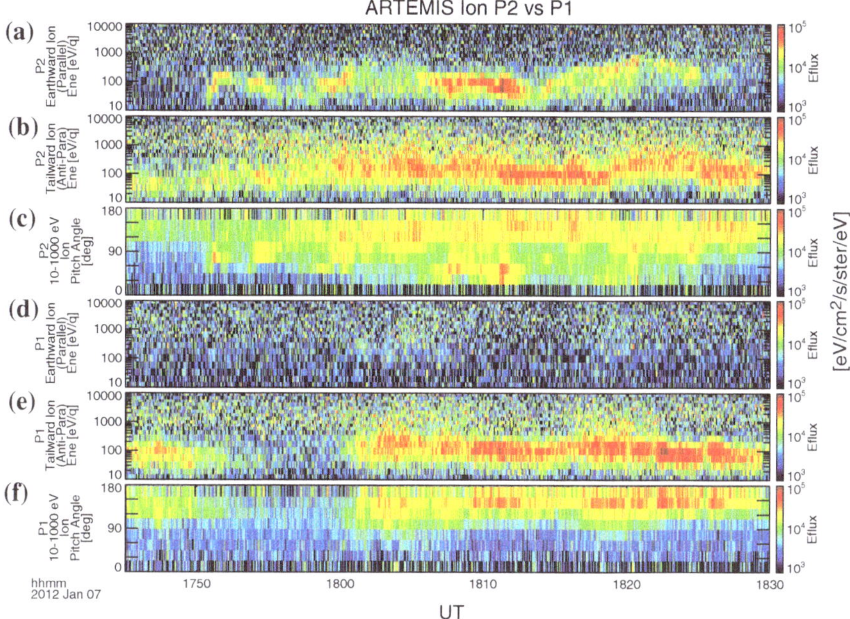

Fig. 3.6 Time-series data of ion energy and pitch-angle spectra from ARTEMIS P2 (**a**–**c**) and P1 (**d**–**f**) during the same time interval as shown in Fig. 3.2. Ion energy spectra in panels (**a**), (**b**), (**d**), and (**e**) are identical to those shown in Figs. 3.2a, b and 3.5a, b respectively

The reduction in the spacecraft potential is presumably caused by an enhancement of the electron current incident onto the spacecraft, which is consistent with the density increase inferred from the plasma-wave observations.

ARTEMIS also observes a Moon-related signature of cold electrons with energies below 100 eV. Figure 3.7 shows a typical electron-energy spectra for different solid angle bins obtained by P2 at 18:10:23–18:10:27. One can see characteristic spectra of photoelectrons emitted from the spacecraft surface and EFI probes in the energy range lower than the measured spacecraft potential. Therefore, it is possible to remove the spacecraft/instrument photoelectrons from the electron-energy spectra by subtracting the spacecraft potential from the electron energies. Figure 3.2d shows the electron energy–time spectrogram from P2, corrected for the spacecraft potential. Here I subtracted the spacecraft potential averaged over the spin from the energy spectrum in units of the distribution function, and converted it to units of the differential energy flux when plotting the corrected energy spectrum. P2 observed a significant cold-electron enhancement at 17:50–17:52, 17:54–18:01, and 18:06–18:15. The electron energy spectra from P1, however, do not show a similar cold-electron enhancement (see Fig. 3.5d), suggesting that the cold electrons detected by P2 are related to the Moon. It is unlikely that the cold-electron enhancement include a significant amount of spacecraft/instrument photoelectrons, because the

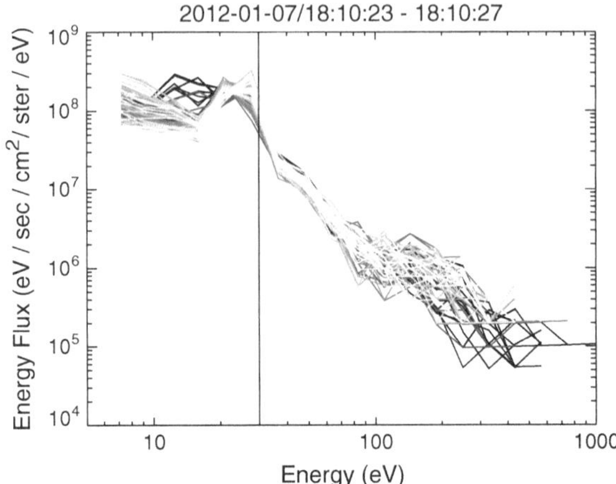

Fig. 3.7 Electron-energy spectra in units of differential energy flux from ARTEMIS P2 at 18:10:23–18:10:27 UT. Energy spectra for the 88 solid angle bins are plotted separately. The vertical black line indicates the spin-averaged spacecraft potential of 30 V. Photoelectrons from the axial and radial EFI sensors are seen as the peaks at 12–16 and 28 eV, respectively [15]. The energy spectrum >30 eV for each bin is clearly different in shape from that of spacecraft/instrument photoelectrons with energies lower than the spacecraft potential, enabling the removal of the photoelectron contamination from the ambient-electron spectrum

energy-spectrum shapes of cold electrons above the spacecraft potential are clearly different from those of the contaminating photoelectrons as shown in Fig. 3.7. One simple interpretation of the cold-electron enhancement measured above the dayside lunar surface is the appearance of newly ionized cold electrons of lunar exospheric origin.

There seem to be good correlations among the Moon-related signatures in the data from P2 described above, i.e., P2 simultaneously observed Earthward-traveling ions, a spacecraft potential reduction, a cold-electron enhancement, and an increase in the electron-plasma frequency at 17:50–17:52, 17:54–18:01, and 18:06–18:15 (see Fig. 3.2a, d, h, and j). The correlations among these Moon-related signatures are consistent with an increase in the numbers of both electrons and ions of lunar exospheric origin ionized above the dayside lunar surface. As I have shown, the dual-probe ARTEMIS observation provides good evidence of the existence of Moon-related plasma with densities comparable to or even several times higher than the ambient-plasma density above the lunar dayside in the terrestrial magnetotail lobes.

3.4 Ambient Electron Modification Associated with Moon-Related Populations

Finally, I point out some interesting signatures in the electron-velocity distributions, which imply interactions between Moon-related plasma populations and ambient electrons of terrestrial and interplanetary origin. P2 observed a depletion in the omnidirectional electron flux at 17:48–17:55 when the field line was not connected to the Moon (see Fig. 3.2c, d). It is easier to see this flux depletion in the pitch-angle distribution panels because of their enhanced color scale (e.g., colors turning from red into yellow at 17:48 in Fig. 3.2g). This overall flux depression during the unconnected interval might represent a convected flux tube that was partially emptied because of lunar shadowing when it was connected to the Moon [1, 11, 13]. In this case, the ARTEMIS P2 on the open field line in the tail lobes should observe Earthward-traveling electrons from the distant tail as soon as the flux tube is detached from the Moon. However, P2 did not detect the Earthward-electron flux ($\alpha < 90°$) larger than the tailward-electron flux ($\alpha > 90°$) in Fig. 3.2e–g. Meanwhile, another Moon-related phenomenon during the unconnected interval was observed, i.e., the spacecraft potential of P2 started to decrease gradually at 17:48 (see Fig. 3.2j). Now I speculate that the Moon-related populations inferred from the spacecraft-potential reduction might be related to the flux depletion during the unconnected interval at 17:48–17:55, though I do not have any plausible explanation for its generation mechanisms yet.

The electron pitch-angle distributions from P2 exhibit intermittent flux depletion around 90° at 17:50–18:22 (see Fig. 3.2e–g). On the other hand, P1 observed almost isotropic electrons throughout this time interval (see Fig. 3.5e–g), indicating that the 90° electron dropouts detected by P2 are Moon-related. The 90° dropouts appear on both the Earthward ($\alpha < 90°$) and tailward ($\alpha > 90°$) sides at 17:50–17:55 when the spacecraft is not connected to the Moon by the field line. During the connected time, dropouts are seen mainly in the downward electron distributions ($\alpha > 90°$), and sometimes in the upward electron distributions ($\alpha < 90°$). The P2 data show good correlations of the 90° electron dropouts with the Moon-related populations discussed in Sect. 3.3, i.e., P2 simultaneously observed the 90° electron dropouts and the signatures of Moon-related populations at 17:50–17:52, 17:54–18:01, and 18:06–18:15 (see Fig. 3.2a, d–h, and j). This suggests direct or indirect interactions of the ambient lobe electrons with the Moon-related populations.

I also note enhanced counter-streaming electrons at 18:25 when the connected interval is just ending (see Fig. 3.2e–g, and the bottom color bar). This is a special geometry, because the field line from the spacecraft passes very close to the Moon. In addition, P2 observed Moon-related ions until 18:25 (see Fig. 3.2a). The combination of these unusual situations might be responsible for generating the enhancement of counter-streaming electrons.

One can see spot-like bright signatures in the upward electron pitch-angle distributions which have almost as high a flux as the incident electrons at 18:02, 18:10–18:13, 18:15, 18:17, and 18:18 (see Fig. 3.2e–g). I presume that they are incident electrons that are magnetically reflected by the crustal fields; for example, the magnetic foot points at 18:10–18:13 are estimated at (lat ~6°N, lon ~12°E), where moderate crustal fields are inferred from electron reflecmetory [9]. They look like "conics" with flux peaks at intermediate pitch angles ($0° < \alpha < 90°$) rather than the loss-cone distributions expected from magnetic reflection of isotropic incident electrons.

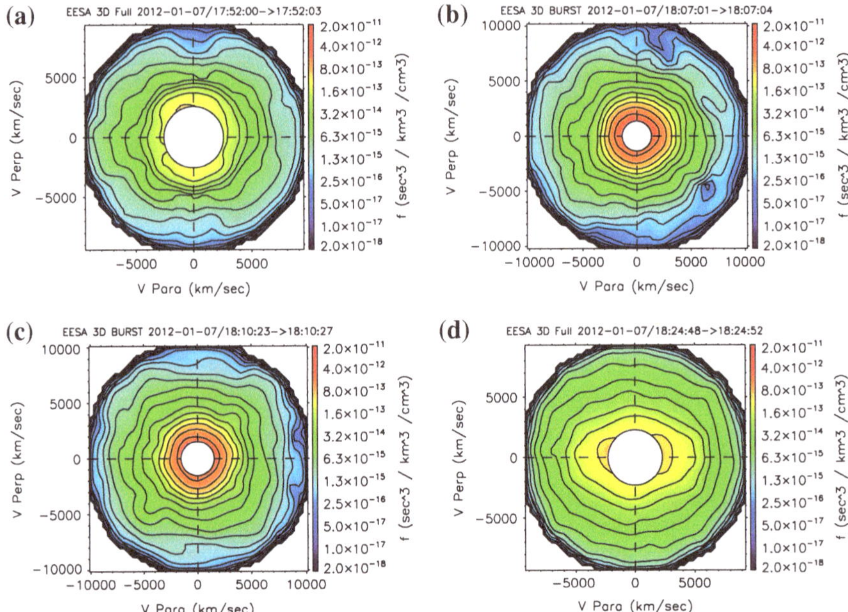

Fig. 3.8 Electron distribution function slices from ARTEMIS P2 as a function of parallel and perpendicular velocity. Positive parallel velocity points Earthward (corresponding to upward motion during the magnetically connected interval). The *inner white circles* indicate the lowest energies of the available data after subtraction of the spacecraft potential

I present in more detail the electron-velocity distributions obtained when the Moon-related signatures were observed. Figure 3.8 shows $v_{||}$–$v_{\perp}$ cuts of the electron-velocity distribution functions obtained at (a) 17:52:00–17:52:03, (b) 18:07:01–18:07:04, (c) 18:10:23–18:10:27, and (d) 18:24:48–18:24:52 (the observed timing is denoted by the black arrows in Fig. 3.2). As seen in the pitch-angle distributions, 90° dropouts are seen in (a) both earthward- and tailward-traveling electron populations, (b) downward-traveling electrons, and (c) both downward- and upward-traveling electrons, with large perpendicular velocities of >5,000 km s^{-1}. The flux depletion of hot electrons with all pitch angles in the positive parallel-velocity region (corresponding to upward electrons) in Fig. 3.8b can be simply interpreted as due to electron absorption by the lunar surface. The upward electron conic in Fig. 3.8c can be interpreted as magnetically reflected anisotropic electrons; if the incident electrons are anisotropic with the 90° dropouts, adiabatic reflection results in an upward-traveling conic. In Fig. 3.8d, counter-streaming electrons are enhanced along the field line without a clear signature of 90° hot-electron dropouts.

Here I briefly discuss the angular distributions of the enhanced cold electrons with energies just above the spacecraft potential. The cold-electron enhancement seen in Fig. 3.8a–c look anisotropic. However, I note that the angular distribution of the lowest-energy electrons cannot be trusted, since the spacecraft potential varies by several volts over the spin period as a result of the differential illumination of the EFI probes. If very cold electrons (e.g., newly ionized electrons of lunar exospheric origin) are present around the spacecraft, they can be strongly modulated by this variable spacecraft potential, resulting in apparent anisotropic enhancement of the lowest-energy electrons. I also note that the corrected energy spectra of the electrons shown in Fig. 3.2d include some smearing effects, because I used the average spacecraft potential over the spin period to apply the spacecraft potential correction.

I now consider possible generation mechanisms for the 90° electron dropouts. To modify the electron-velocity distribution functions, electric or magnetic fields, or source/loss processes are needed. As a way of removing the 90° electrons, I can consider collisions, wave–particle interactions ("microscopic" electric and magnetic fields), and large-scale electrostatic and magnetic structures ("macroscopic" electric and magnetic fields). However, my evaluation shows that the cross sections for collisions are too small for any reasonable density of target particles, such as neutral atoms and dust grains, to achieve the observed erosion of ambient electrons. Typical cross sections of neutral atoms $\sigma_{\mathrm{en}} < 10^{-19}\,\mathrm{m}^2$ [10] and number densities near the surface $n_{\mathrm{n}} < 10^6\,\mathrm{cm}^{-3}$ [24] yield mean free paths $\lambda_{\mathrm{en}} > 10^4\,\mathrm{km}$, which is much longer than the lunar radius of 1,738 km. Besides, the mean free path will increase rapidly, because the number density falls off exponentially as a function of altitude. Therefore, I would not expect that collisions with neutral atoms can take place sufficiently frequently to remove electrons. As for charged dust grains that are potentially levitating around the Moon [25], if I assume that a dust particle with a radius $R_{\mathrm{d}} \sim 0.1\,\mu\mathrm{m}$ is charged to $\phi_{\mathrm{d}} \sim +10\,\mathrm{V}$, I can calculate its capacitance, $C \sim 4\pi\varepsilon_0 R_{\mathrm{d}} \sim 10^{-17}\,\mathrm{F}$ [6], and charge, $q = C\phi_{\mathrm{d}} \sim 10^{-16}\,\mathrm{C}$. Considering Coulomb collisions, its cross section for 100 eV incident electrons is calculated as $\sigma_{\mathrm{ed}} < 10^{-14}\,\mathrm{m}^2$. Using expected dust densities, n_{d}, of 10^{-5} to $10^{-6}\,\mathrm{cm}^{-3}$ at 100–200 km altitude [12], the mean free path will be $\lambda_{\mathrm{ed}} > 10^{10}\,\mathrm{km}$, which is again much longer than the size of the system.

Next I show the electric and magnetic field wave data. The relative motion between the newborn populations and the background plasma flow would provide the free energy source for exciting waves like other environments including comets and Mars, where various waves associated with pickup ions have been observed [23, 27]. The excited wave fields then might modify the electron velocity distribution functions. Figure 3.9b–g shows electric and magnetic field wave spectra from P2 and P1. The high frequency magnetic field spectra are obtained by the SCM (Search Coil Magnetometer) instruments [22]. The magnetic field spectra are very flat with no indication of electromagnetic waves such as whistlers (Fig. 3.9c, f). The wavelet spectra of low frequency magnetic field also shows no clear difference between the data from the two probes (Fig. 3.9d, g). Thus ARTEMIS did not observe any Moon-related electromagnetic wave excitations with frequencies from above the electron cyclotron frequency down to near-DC level.

One might expect excitations of electrostatic waves in the low-β plasma in the tail lobe. The P2 data show some electrostatic noise with frequencies of ~10–100 Hz (Fig. 3.9b), but it does not seem to be correlated with the 90° electron dropouts (Fig. 3.2e–g). Besides, since P1 occasionally observed similar enhancements of the electric field wave intensity with frequencies of 10–100 Hz (Fig. 3.9e), I would not conclude that the electrostatic noise detected by P2 is the Moon-related signature

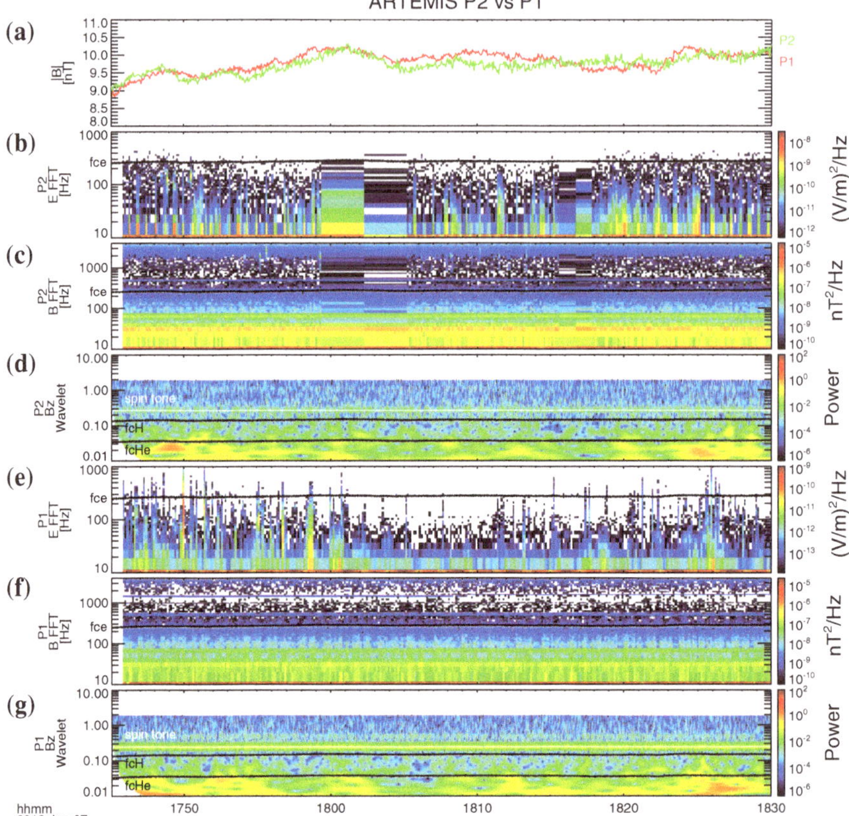

Fig. 3.9 Comparison of the data from ARTEMIS P1 and P2 on 7 January 2012. **a** Magnetic field intensities from P1 (*red*) and P2 (*green*), FFT wave spectra of high frequency **b** electric and **c** magnetic field, **d** wavelet spectra of low frequency magnetic field from P2, and (**e**–**g**) from P1

responsible for generating the 90° electron dropouts. Thus, the wave data from ARTEMIS do not seem to be favorable to the wave-particle interaction hypothesis. Therefore, hereafter I explore the possibility of the presence of large-scale electrostatic and magnetic structures.

Both large-scale electrostatic and magnetic structures can produce 90° electron dropouts. If a field-aligned potential structure exists, as shown in Fig. 3.10a, incident electrons will be accelerated by the electric field along the magnetic field line. Electrons with initially zero parallel-velocity component will have a parallel-velocity component of $\sqrt{2e\phi/m}$ after having been accelerated, where e is the elementary charge, ϕ the field-aligned potential difference, and m the electron mass. The field-aligned acceleration cases a gap in velocity space with a constant cutoff parallel velocity $v_{||c} = \sqrt{2e\phi/m}$. I note that the electrostatic potential that I postulate here is probably not related to the lunar surface potential, because the surface potential

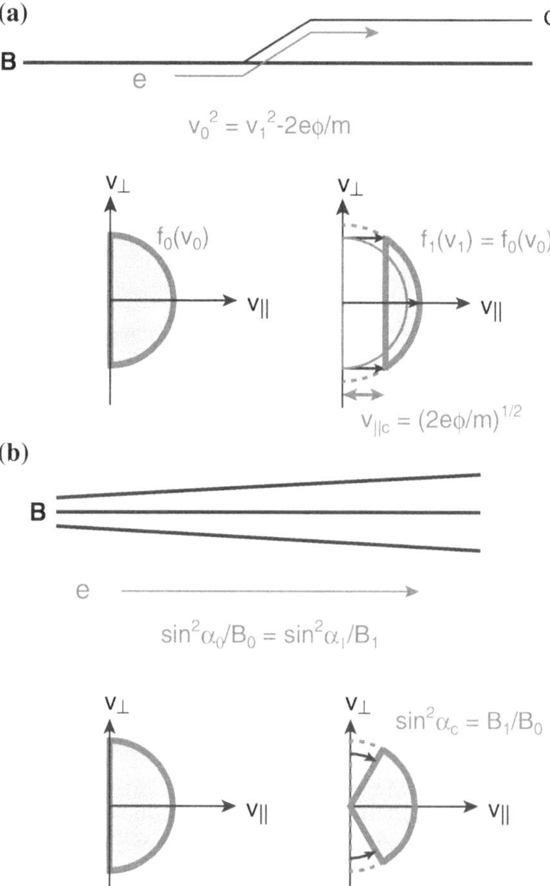

Fig. 3.10 Schematic illustration of possible generation mechanisms of the 90° electron dropouts associated with **a** electrostatic acceleration along the magnetic field and **b** a pitch-angle focusing effect owing to a nonuniform magnetic-field configuration. The lunar surface and the spacecraft are located to the right along the magnetic-field line. Initially isotropic downward-traveling electrons are modified by the time they reach the spacecraft. Note that the electrostatic process produces energy-dependent cutoff pitch angles, whereas the magnetic process results in energy-independent cutoff pitch angles

should be shielded within a few Debye lengths, which is much smaller than the spacecraft altitude. On the other hand, if a nonuniform magnetic-field configuration exists, as shown in Fig. 3.10b, conservation of the first invariant $\mu = mv_\perp^2/2B$ will render the electron pitch angles smaller. This pitch-angle focusing effect will also produce a gap in velocity space around $\alpha = 90°$, but with a constant cutoff pitch angle of $\alpha_c = \sin^{-1}\sqrt{B_1/B_0}$, where B_1 and B_0 are the magnetic-field strengths at the spacecraft and in the upside region, respectively ($B_0 > B_1$). Therefore, if an electrostatic process is responsible for the 90° electron dropouts, the cutoff pitch angles will be energy-dependent, whereas the magnetic process will give rise to energy-independent cutoff pitch angles.

Here I examine the energy dependence of the electron dropout angles around 90° from the P2 data. I define the cutoff pitch angle at each energy as the angle for which the slope of the electron pitch-angle distribution, $s = \partial f/\partial \alpha$, becomes maximal for $90° < \alpha < 180°$ and minimal for $0° < \alpha < 90°$. I exclude upward-traveling electrons during the magnetically connected interval, adopt thresholds for the slope to get rid of flat isotropic distributions, and then plot the remaining cutoff pitch angles using red, blue, and black lines in Fig. 3.2e–g. The cutoff pitch angles for three energies are also plotted together in Fig. 3.2k. Rapid fluctuations are seen in the cutoff pitch angle for 289 eV, but the three cutoff pitch angles seem to behave in a similar fashion, suggesting that they are energy-independent. For field-aligned electrostatic acceleration with a constant cutoff parallel velocity, the cutoff pitch angles for different electron energies will obey a relationship of the form $\sqrt{E_1} \cos \alpha_{c1} = \sqrt{E_2} \cos \alpha_{c2}$, where α_{c1} and α_{c2} are the cutoff pitch angles for electron energies E_1 and E_2, respectively. This relationship is obviously not satisfied most of the time (cf. Fig. 3.2k). Thus, the lack of energy dependence of the cutoff pitch angles implies a magnetic rather than an electrostatic origin of the 90° electron dropouts. I note, despite the implication of nonuniform magnetic fields suggested by the electron data, that the magnetic-field data from the two probes indicate almost identical magnetic fields (see Figs. 3.5i and 3.9a). I will leave a detailed discussion of these enigmatic characteristics of the electron modification to a future investigation.

As I have described in this section, the ARTEMIS data exhibit very curious features of electron-velocity distribution functions associated with plasma of lunar origin. Note that the downward electrons of terrestrial or interplanetary origin are already modified by the time they reach the spacecraft and the lunar surface. This means that the interaction region of ambient electrons and Moon-related populations extends at least several hundred kilometers above the dayside lunar surface.

3.5 Conclusions

ARTEMIS observed field-aligned, upward-traveling electrons from the dayside lunar surface when the Moon was located in the terrestrial magnetotail lobes. Their pitch-angle distributions are in good agreement with the theoretical predictions of gyrophase-limited electron emission from an extended surface in an oblique magnetic field. The energy distributions of the upward electrons comprise a large population of over 100 % of the incident electrons with energies in the range ~10–50 eV and 30 % with energies of ~50–200 eV. I speculate that at least the lower-energy population of the upward-traveling electrons primarily consist of the high-energy tail of photoelectrons emitted from the lunar surface. These new observations help resolve long-standing questions about the existence of large dayside surface positive potentials first inferred from Apollo Charged Particle Lunar Environment Experiment (CPLEE) measurements. The large upward flux with relatively high energies of ~10–200 eV implies large positive potentials of the lunar surface. The large positive potentials will accelerate a significant amount of newly ionized ions of lunar origin near the lunar surface, both parallel and perpendicular to the magnetic field, and play an important role in the plasma dynamics around the Moon in the tail lobe.

The dual-probe ARTEMIS mission has revealed Moon-related signatures of ions and electrons in the tail lobes, which are only detected when the probe passes above the Moon's dayside. The lunar dayside probe simultaneously observes Moon-related ions, an increase in the electron plasma frequency, a spacecraft potential reduction, and a cold-electron enhancement. The Moon-related Earthward ions have both parallel- and perpendicular-velocity components. The nonzero parallel-velocity component of the pickup ions suggests large positive potentials of tens of volts on the lunar surface, supporting the existence of a high-energy tail of lunar-surface photoelectrons. The electric-field wave spectra from the two probes show that the electron-plasma frequency rises during the lunar dayside flyby. A comparison of the electron-plasma frequencies between the two probes indicates the presence of several times as many ions and electrons above the dayside lunar surface as in the ambient plasma. The Moon-related spacecraft potential reduction indicates an enhancement of the electron current incident onto the spacecraft, which also implies that the plasma density increases. The electron data show cold-electron enhancements with energies below $\sim$100 eV, which suggest the existence of newly ionized cold electrons of lunar exospheric origin. The coexistence of these Moon-related signatures is consistent with increases in the numbers of both electrons and ions ionized above the dayside lunar surface. Thus, the dual-probe ARTEMIS measurements confirm that plasma of lunar origin dominates the terrestrial and interplanetary plasma populations above the Moon's dayside in the Earth's magnetotail lobes.

The ARTEMIS data also exhibit modified electron-velocity distribution functions that coexist with the Moon-related populations. The electron data show 90° electron dropouts, enhanced counter-streaming electrons, and upward electron conics. The upward-traveling electron conics can be interpreted as magnetically reflected anisotropic electrons with 90° dropouts. The good correlations of the 90° electron dropouts with the signatures of plasma of lunar origin imply direct or indirect interactions of ambient lobe electrons with lunar pickup ions and electrons. The lack of energy dependence of the 90° dropouts implies a magnetic origin, whereas the magnetometer measurements by the two probes show almost identical magnetic fields. In future work, I will investigate in more detail the generation mechanisms of the 90° electron dropouts and counter-streaming electrons.

The comprehensive observations of the lunar-dayside plasma environment in the tail lobes conducted by the dual-probe ARTEMIS mission provide us with a wealth of information about Moon-related ion and electron populations and their interactions with the ambient plasma. I have shown that electrons and ions of both lunar and ambient origins are closely connected. Both the incident electron flux from the ambient plasma and the energy distribution of the upward-traveling electrons primarily decide the lunar surface potential. The large positive potential of the lunar surface, in turn, accelerates the dense newly ionized heavy ions outward from the Moon's dayside. These Moon-related ions potentially interact with the ambient plasma in the tail lobes. In this environment, the heavy ions and cold electrons of lunar exospheric origin can predominantly exist in an ambient, low-β plasma which is characterized by a sub-Alfvénic flow speed. Further analysis of the ARTEMIS data, as well as theoretical work and numerical simulations, will be necessary to completely understand the observed characteristics and multiple-species plasma dynamics.

References

1. Anderson, K.A., Lin, R.P.: Observation of interplanetary field lines in the magnetotail. J. Geophys. Res. **74**(16), 3953–3968 (1969). doi:10.1029/JA074i016p03953, http://dx.doi.org/10.1029/JA074i016p03953
2. Anderson, K.A., Lin, R.P., McGuire, R.E., McCoy, J.E.: Measurement of lunar and planetary magnetic fields by reflection of low energy electrons. Space Sci. Instrum. **1**, 439–470 (1975)
3. Angelopoulos, V.: The ARTEMIS mission. Space Sci. Rev. **165**(1–4), 3–25 (2011). doi:10.1007/s11214-010-9687-2
4. Auster, H.U., Glassmeier, K.H., Magnes, W., Aydogar, O., Baumjohann, W., Constantinescu, D., Fischer, D., Fornacon, K.H., Georgescu, E., Harvey, P., Hillenmaier, O., Kroth, R., Ludlam, M., Narita, Y., Nakamura, R., Okrafka, K., Plaschke, F., Richter, I., Schwarzl, H., Stoll, B., Valavanoglou, A., Wiedemann, M.: The THEMIS fluxgate magnetometer. Space Sci. Rev. **141**(1–4), 235–264 (2008). doi:10.1007/s11214-008-9365-9
5. Bonnell, J.W., Mozer, F.S., Delory, G.T., Hull, A.J., Ergun, R.E., Cully, C.M., Angelopoulos, V., Harvey, P.R.: The electric field instrument (EFI) for THEMIS. Space Sci. Rev. **141**(1–4), 303–341 (2008). doi:10.1007/s11214-008-9469-2
6. Goertz, C.K.: Dusty plasmas in the solar system. Rev. Geophys. **27**(2), 271–292 (1989). doi:10.1029/RG027i002p00271
7. Halekas, J.S., Delory, G.T., Farrell, W.M., Angelopoulos, V., McFadden, J.P., Bonnell, J.W., Fillingim, M.O., Plaschke, F.: First remote measurements of lunar surface charging from ARTEMIS: evidence for nonmonotonic sheath potentials above the dayside surface. J. Geophys. Res. **116**, A07103 (2011). doi:10.1029/2011JA016542
8. Halekas, J.S., Lillis, R.J., Lin, R.P., Manga, M., Purucker, M.E., Carley, R.A.: How strong are lunar crustal magnetic fields at the surface?: considerations from a reexamination of the electron reflectometry technique. J. Geophys. Res. **115**, E03006 (2010). doi:10.1029/2009JE003516
9. Halekas, J.S., Mitchell, D.L., Lin, R.P., Frey, S., Hood, L.L., Acuña, M.H., Binder, A.B.: Mapping of crustal magnetic anomalies on the lunar near side by the Lunar prospector electron reflectometer. J. Geophys. Res. **106**, 27841–27852 (2001). doi:10.1029/2000JE001380
10. Kieffer, L.J., Dunn, G.H.: Electron impact ionization cross-section data for atoms, atomic ions, and diatomic molecules: I. experimental data. Rev. Mod. Phys. **38**, 1–35 (1966). doi:10.1103/RevModPhys.38.1, http://link.aps.org/doi/10.1103/RevModPhys.38.1
11. Lin, R.P.: Observations of Lunar shadowing of energetic particles. J. Geophys. Res. **73**, 3066–3071 (1968). doi:10.1029/JA073i009p03066
12. McCoy, J.E., Criswell, D.R.: Evidence for a high altitude distribution of lunar dust. In: Proceedings of Lunar sciience conference, 5th, pp. 2991–3005 (1974)
13. McCoy, J.E., Lin, R.P., McGuire, R.E., Chase, L.M., Anderson, K.A.: Magnetotail electric fields observed from Lunar orbit. J. Geophys. Res. **80**, 3217–3224 (1975). doi:10.1029/JA080i022p03217
14. McFadden, J.P., Carlson, C.W., Larson, D., Bonnell, J., Mozer, F., Angelopoulos, V., Glassmeier, K.H., Auster, U.: THEMIS ESA first science results and performance issues. Space Sci. Rev. **141**(1–4), 477–508 (2008). doi:10.1007/s11214-008-9433-1
15. McFadden, J.P., Carlson, C.W., Larson, D., Ludlam, M., Abiad, R., Elliott, B., Turin, P., Marckwordt, M., Angelopoulos, V.: The THEMIS ESA plasma instrument and in-flight calibration. Space Sci. Rev. **141**(1–4), 277–302 (2008). doi:10.1007/s11214-008-9440-2

16. Poppe, A., Halekas, J.S., Horányi, M.: Negative potentials above the day-side lunar surface in the terrestrial plasma sheet: evidence of non-monotonic potentials. Geophys. Res. Lett. **38**, L02103 (2011). doi:10.1029/2010GL046119
17. Poppe, A., Horányi, M.: Simulations of the photoelectron sheath and dust levitation on the lunar surface. J. Geophys. Res. **115**, A08106 (2010). doi:10.1029/2010JA015286
18. Poppe, A.R., Halekas, J.S., Delory, G.T., Farrell, W.M., Angelopoulos, V., McFadden, J.P., Bonnell, J.W., Ergun, R.E.: A comparison of ARTEMIS observations and particle-in-cell modeling of the lunar photoelectron sheath in the terrestrial magnetotail. Geophys. Res. Lett. **39**, L01102 (2012). doi:10.1029/2011GL050321
19. Poppe, A.R., Halekas, J.S., Samad, R., Sarantos, M., Delory, G.T.: Model-based constraints on the lunar exosphere derived from ARTEMIS pickup ion observations in the terrestrial magnetotail. J. Geophys. Res. **118**(5), 1135–1147 (2013). doi:10.1002/jgre.20090, http://dx.doi.org/10.1002/jgre.20090
20. Poppe, A.R., Samad, R., Halekas, J.S., Sarantos, M., Delory, G.T., Farrell, W.M., Angelopoulos, V., McFadden, J.P.: ARTEMIS observations of lunar pick-up ions in the terrestrial magnetotail lobes. Geophys. Res. Lett. **39**, L17104 (2012). doi:10.1029/2012GL052909
21. Reiff, P.H.: Magnetic shadowing of charged particles by an extended surface. J. Geophys. Res. **81**(19), 3423–3427 (1976). doi:10.1029/JA081i019p03423
22. Roux, A., Le Contel, O., Coillot, C., Bouabdellah, A., de la Porte, B., Alison, D., Ruocco, S., Vassal, M.C.: The search coil magnetometer for THEMIS. Space Sci. Rev. **141**(1–4), 265–275 (2008). doi:10.1007/s11214-008-9455-8
23. Sauer, K., Dubinin, E., Baumgartel, K., Tarasov, V.: Low-frequency electromagnetic waves and instabilities within the Martian bi-ion plasma. Earth Planets Space **50**(3), 269–278 (1998)
24. Stern, S.: The lunar atmosphere: History, status, current problems, and context. Rev. Geophys. **37**(4), 453–491 (1999). doi:10.1029/1999RG900005
25. Stubbs, T., Vondrak, R., Farrell, W.: A dynamic fountain model for lunar dust. Adv. Space Res. **37**(1), 59–66 (2006). doi:10.1016/j.asr.2005.04.048
26. Tanaka, T., Saito, Y., Yokota, S., Asamura, K., Nishino, M.N., Tsunakawa, H., Shibuya, H., Matsushima, M., Shimizu, H., Takahashi, F., Fujimoto, M., Mukai, T., Terasawa, T.: First in situ observation of the Moon-originating ions in the Earth's Magnetosphere by MAP-PACE on SELENE (KAGUYA). Geophys. Res. Lett. **36**, L22106 (2009). doi:10.1029/2009GL040682
27. Tsurutani, B.T.: Comets: a laboratory for plasma waves and instabilities. Geophys. Monogr. Ser. **61**, 189–209 (1991). doi:10.1029/GM061p0189
28. Whipple, E.C.: Potentials of Surfaces in Space. Rep. Prog. Phys. **44**(11), 1197–1250 (1981). doi:10.1088/0034-4885/44/11/002, http://stacks.iop.org/0034-4885/44/i=11/a=002
29. Zhou, X.Z., Angelopoulos, V., Poppe, A.R., Halekas, J.S.: Artemis observations of lunar pick-up ions: mass constraints on ion species. J. Geophys. Res. **118**, 1766–1774 (2013). doi:10.1002/jgre.20125, http://dx.doi.org/10.1002/jgre.20125

Chapter 4
Hot-Proton Interactions with the Surface and Magnetic Anomalies of the Moon

Abstract Proton backscattering at the lunar surface is an important consequence of the direct interaction of the ambient plasma and the Moon. Furthermore, observations of the backscattered energetic neutral atoms (ENAs) can be used as a remote sensing technique to monitor the incident proton flux to the surface and to investigate the lunar surface shielding by crustal magnetization. This chapter describes the characteristics of the backscattered hydrogen ENAs observed by the lunar-orbiting Chandrayaan-1 spacecraft in the terrestrial plasma sheet. The energy distributions of the hydrogen ENAs in the plasma sheet are well-represented by the Maxwellian with the best-fit temperatures of $\sim$100 eV, which are similar to those of the backscattered hydrogen ENAs detected in the solar wind. On the other hand, comparison of the hydrogen ENA/proton flux ratios obtained in the plasma sheet and in the solar wind suggests less effective magnetic shielding in the plasma sheet. Test-particle simulations demonstrate that the broad velocity distributions of the plasma sheet protons prevent development of a distinct mini-magnetosphere, which is formed in the beam-like solar wind protons.

Keywords Energetic neutral atoms · Backscattering · Terrestrial plasma sheet · Lunar mini-magnetosphere

4.1 Introduction

Up to this point, I have mainly discussed the electron dynamics both near and far above the lunar surface. This chapter shows lunar energetic neural atom (ENA) observations within the Earth's magnetotail and describes the ion motion around lunar crustal magnetic fields. Since the ion gyroradius is much larger than the electron gyroradius, ions will behave non-adiabatically even at the largest magnetic anomalies on the Moon. Taking into account both ion and electron dynamics is critical in fully understanding plasma interactions with crustal magnetic fields. In this chapter, I study the ion motion around lunar magnetic anomalies from ENA measurements as well as from test-particle simulations.

Y. Harada, *Interactions of Earth's Magnetotail Plasma with the Surface, Plasma, and Magnetic Anomalies of the Moon*, Springer Theses,
DOI 10.1007/978-4-431-55084-6_4

As introduced in Chap. 1, all of the reported lunar ENA observations were conducted when the Moon was located in the solar wind or magnetosheath. The Moon enters the Earth's magnetotail for several days around full moon. In the magnetotail, the Moon is sometimes exposed to hot plasma of the terrestrial plasma sheet [4, 14]. Plasma parameters in the plasma sheet such as densities, ion and electron temperatures and Mach numbers significantly differ from those in the solar wind. Therefore, lunar ENA characteristics in the Earth's magnetotail can be quite different from those in the solar wind. The similarity and difference is directly associated with nature of the plasma interaction with the regolith surface in the two different plasma regimes. Here I present ENA, plasma and magnetic-field data from Chandrayaan-1 and Kaguya when the Moon was located within the terrestrial magnetotail, and discuss the lunar ENA characteristics in the plasma sheet.

4.2 Instrumentation and Datasets

In this chapter, I mainly analyze the data obtained by the Sub-keV Atom Reflecting Analyzer (SARA) instrument onboard Chandrayaan-1, which consists of two sensors: the Solar Wind Monitor (SWIM) and the Chandrayaan-1 Energetic Neutrals Analyzer (CENA) [1]. SWIM measures ions ($\sim$100 eV/q to 3 keV/q) with mass resolution within a $\sim$7° × 160° field of view (FOV) divided into 16 angular sectors [11]. CENA measures energetic neutral atoms in the energy range from $\sim$10 to 3 keV with mass resolution within a 9° × 160° FOV divided into seven angular sectors [8]. In this paper, I use the CENA data in the energy range of 38–652 eV (energy setting 2) in the hydrogen channels from three angular sectors centered at the nadir-looking direction (sectors 2, 3 and 4).

Some of the CENA data obtained in the sunlight are contaminated by ultraviolet light, and sometimes it is difficult to extract the rather weak ENA signal in the magnetotail. Note that the ENA signal in the magnetotail is weaker by 2–3 orders of magnitude compared with that in the solar wind. Because of this difficulty, here I only use the CENA data obtained when the Chandrayaan-1 was in the optical shadow of the Moon. Nevertheless, plasma sheet ions can even precipitate onto the Moon's nightside because both earthward (corresponding to the sunward motion) and tailward (anti-sunward) plasma flows exist at the lunar distance ($\sim$60R_E, where R_E is the Earth's radius of 6,378 km) in the Earth's magnetotail [12, 16, 21]. Besides, hot protons with nearly isotopic velocity distributions are frequently observed in the central plasma sheet [9]. Thus, the nightside lunar surface is exposed to incident protons when the Moon is immersed in the isotropic plasma.

I also present the plasma and magnetic-field data from the MAgnetic field and Plasma experiment (MAP) instrument onboard Kaguya to confirm the ambient plasma conditions around the Moon at the time when the SARA data were obtained. As described in Sect. 2.2, MAP consists of two components: Lunar MAGnetometer (LMAG) and Plasma energy Angle and Composition Experiment (PACE). LMAG measures the magnetic field with a resolution of 0.1 nT [10, 19, 20, 22]. PACE consists of four particle sensors: two electron spectrum analyzers (ESA-S1 and ESA-S2), an ion mass analyzer (IMA), and an ion energy analyzer (IEA) [17, 18]. Each sensor has a hemispherical FOV to obtain three-dimensional velocity distribution functions of electrons and ions.

4.3 ENAs from the Moon in the Earth's Magnetotail

4.3.1 Overview of Lunar ENA Observations in the Earth's Magnetotail

I first present time-series data of ENAs and ambient-plasma conditions from two time intervals when the Moon was in the geomagnetic tail and Chandrayaan-1 was in the optical shadow of the Moon. The spacecraft trajectories during these time intervals are shown in Fig. 4.1. The selenocentric solar ecliptic (SSE) system has its x-axis from the Moon toward the Sun, the z-axis is parallel to the upward normal to the Earth's ecliptic plane, and the y completes the orthogonal coordinate set, whereas the geocentric solar ecliptic (GSE) system has its x-axis from the Earth toward the Sun. During these orbits, the Moon was located in the dawn-side magnetotail (~01–02 LT). Figures 4.2 and 4.3 show the SARA data for orbits 2565 and 2555, respectively, as well as the ion, electron, and magnetic-field data simultaneously obtained by MAP onboard Kaguya. In orbit 2565, Chandrayaan-1 traveled from north to south

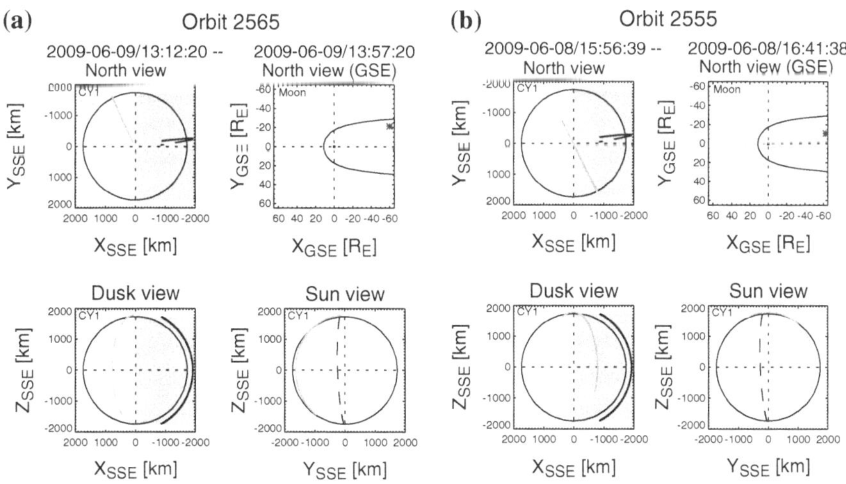

Fig. 4.1 Chandrayaan-1 (CY-1) and Kaguya (KGY) trajectories for CY-1 orbits **a** 2565 and **b** 2555 in SSE coordinates and the position of the Moon in GSE coordinates (R_E = 6,378 km, the *gray zones* in the *left panels* represent the optical shadow of the Moon and the *black lines* in the *top right panels* show the typical location of the magnetopause)

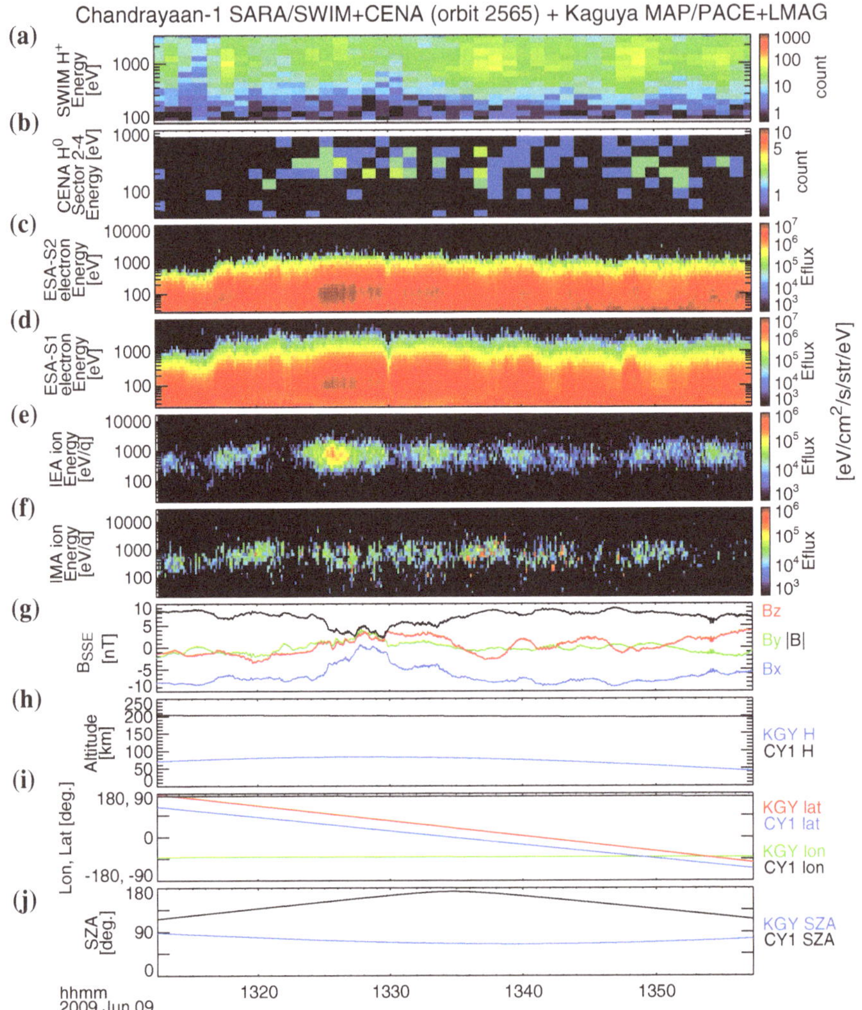

Fig. 4.2 Time series data from Chandrayaan-1 (orbit 2565) and Kaguya in the terrestrial plasma sheet. Energy-time spectrograms of **a** protons from CY-1/SWIM (64 s data accumulation), **b** hydrogen atoms from CY-1/CENA (64 s data accumulation), **c** downward electrons from KGY/ESA-S1, **d** upward electrons from KGY/ESA-S2, **e** downward ions from KGY/IEA, **f** upward ions from KGY/IMA, **g** magnetic field in SSE coordinates from KGY/LMAG, spacecraft **h** altitudes, **i** selenographic longitudes and latitudes, and **j** solar zenith angles are shown

at ~200 km altitudes at selenographic longitudes of ~170°E, while Kaguya was at <100 km altitudes on the near-terminator orbit (Fig. 4.2h, i). Figure 4.4 displays the Chandrayaan-1 orbit track and surface area on the lunar surface seen by the CENA angular sectors 2–4. The FOV projected area in the southern hemisphere corresponds to the large magnetic anomaly, where the ENA backscattering ratio map previously

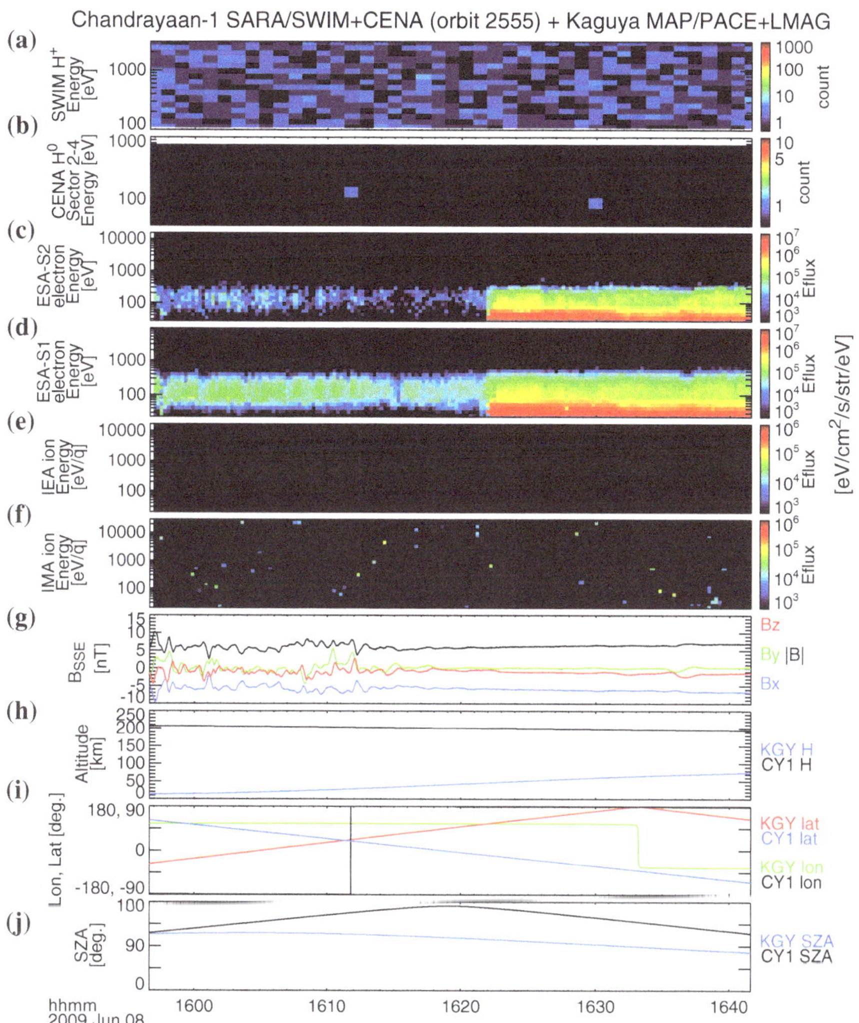

Fig. 4.3 Time series data from Chandrayaan-1 (orbit 2555) and Kaguya in the terrestrial magnetotail lobe in the same format as in Fig. 4.2. Kaguya was illuminated by sunlight after 16:22 UT and intense spacecraft photoelectrons which were attracted by the positive spacecraft potential can be seen in the low-energy part of the **d** ESA-S1 and **c** ESA-S2 data. The fluctuations observed by **g** LMAG at 15:56–16:18 UT at low altitudes are presumably due to lunar crustal magnetic fields

obtained from the CENA data in the solar wind exhibits a significant reduction in ENA backscattering ratio, with reductions up to ~50 % [25]. Meanwhile, Kaguya observed hot electrons (Fig. 4.2c, d) and ions (Fig. 4.2e, f) with energies of hundreds eV up to a few keV, indicating that the Moon was located in the terrestrial plasma sheet. The SWIM data also exhibit consistent hot proton signatures (Fig. 4.2a), which truly show that the Moon was exposed to nearly isotropic flux of the plasma sheet

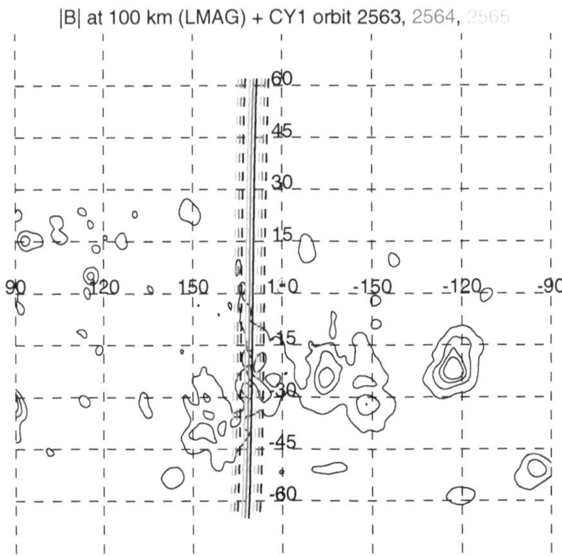

Fig. 4.4 Selenographic projection of the Chandrayaan-1 trajectories in the optical shadow of the Moon during the orbits 2563 (*black, solid, thick line*), 2564 (*dark*), and 2565 (*light gray*). The *dashed thick lines* represent the surface projections of the field of view of the CENA sectors 2–4. The *black thin contours* indicate the lunar crustal magnetic field strength at 100 km altitude obtained from KGY/LMAG data with lines for 0.5, 1, 1.5, and 2 nT [22, 23]

protons. On the other hand, the plasma data from orbit 2555 (Fig. 4.3a, c–f) show the absence of hot electrons and ions, indicating that the Moon was located in the tail lobe.

Simultaneously when SWIM identified the plasma sheet protons, CENA detected hydrogen ENAs upcoming from the Moon. Figures 4.2b and 4.3b show the energy-time spectrograms of hydrogen ENAs from the CENA sectors 2–4. The CENA counts are seen during most of the time of orbit 2565. These continuous signals are absent in Fig. 4.3b, which only shows a few instrument background counts in the optical shadow. The correlation of the CENA counts with the proton flux detected by SWIM implies that the detected CENA counts are not caused by background noise.

4.3.2 ENA Energy Spectrum

Generation mechanisms of the ENAs from the Moon can be deduced from their energy spectra. Figure 4.5 shows the energy spectrum of hydrogen ENAs from CENA in the optical shadow of the Moon, integrated from orbits 2563 (08:57:12–09:42:12 UT), 2564 (11:04:46–11:49:46 UT), and 2565 (13:12:20–13:57:20 UT) on 9 Jun 2009. The integration time for each orbit is ~45 min, resulting in a total accumulation

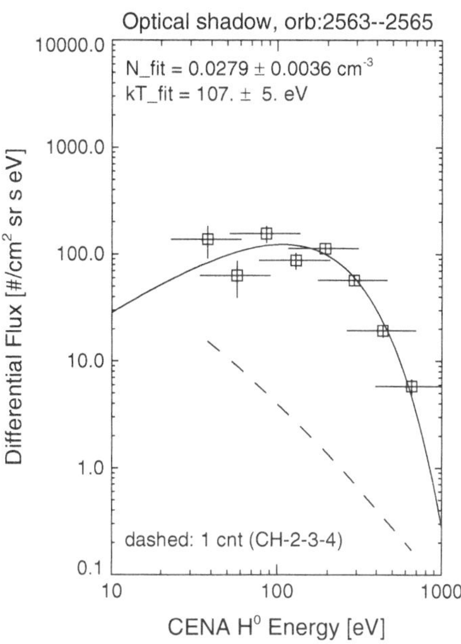

Fig. 4.5 Energy spectrum of upward-traveling hydrogen ENAs detected by CENA in the optical shadow of the Moon integrated over the orbits 2563–2565. Error bars correspond to the energy resolution (x-axis) and the uncertainty due to counting statistics (y-axis). The *solid curve* shows the best-fitted Maxwell distribution. The fitting parameters with errors from propagation of the fitting errors are also denoted. The *dashed line* represents the one count level (over 135 min)

time of ~135 min. During most of these time periods, the Moon was located in the Earth's plasma sheet and SWIM observed significant proton flux. The ENA energy spectrum obtained in the plasma sheet has a similar shape to the energy spectra of backscattered hydrogen ENAs observed by IBEX and Chandrayaan-1 in the solar wind [2, 15]. The ENA energy spectra observed by CENA are well-reproduced by the Maxwell distribution with typical characteristic energies of ~60–140 eV [2]. The solid curve in Fig. 4.5 shows the best-fitted Maxwell distribution (the observed data are weighted by the uncertainty due to counting statistics when fitting). One can see that the observed spectrum in the plasma sheet is well-fitted by the Maxwell distribution especially in the higher energy range with sufficient statistics. The best fit parameters are the number density $N \sim 0.03\,\mathrm{cm}^{-3}$ and the characteristic energy (the e-folding energy of the distribution function) $kT \sim 100\,\mathrm{eV}$. The characteristic energy of ENAs in the plasma sheet is comparable to those in the solar wind. As discussed in the previous paper [2], the shape of these energy spectra is not compatible with the ENA energy spectra produced by ion sputtering, but can be explained by backscattering of ions from the surface of regolith grains, typically involving a few scatter events for a backscattered ENA. Thus, it is most likely that the detected ENAs from the Moon in the plasma sheet are also backscattered hydrogen ENAs.

As in the previous work using CENA observations in the solar wind [2], I use the Maxwell distribution only because it best fits the observed ENA energy spectra. However, this is just an empirically found spectrum and not derived from the physics of multiple scattering events at the lunar surface. Also note that since the collisionless ENAs are not in thermal equilibrium, the best-fit parameters derived here do not represent the density and temperature of a thermal gas. Physical explanation and detailed discussion of the backscattering processes at the lunar surface are beyond the scope of this thesis.

4.3.3 Backscattering Ratio

Here I examine the flux ratio of the backscattered hydrogen ENAs to the incident protons. I define the backscattering ratio as

$$\begin{aligned} r &= \frac{F_{xH^0}}{F_{xH^+}} \\ &= \frac{|\iint \cos\theta J_{H^0}(E,\Omega)dEd\Omega|}{|\iint \cos\theta J_{H^+}(E,\Omega)dEd\Omega|}, \end{aligned} \tag{4.1}$$

where F_{xH^0} is the upward flux of the hydrogen ENAs from the surface, F_{xH^+} is the downward flux of the incident protons onto the surface, θ is the angle between the particle velocity and the surface normal, J_{H^0} and J_{H^+} are the differential number fluxes of the upward hydrogen ENAs and downward protons, respectively. I assume an isotropic angular distribution of the observed hydrogen ENAs for simplicity as was done in the previous paper [2]. The actual angular distribution is not too different form the isotropic one [25].

I adopt two different approaches to calculate the ENA and proton fluxes: using Maxwellian best-fit parameters [2] and directly integrating the observed differential flux over the measured energy range [15]. Using the best-fit parameters, the upward ENA flux is calculated as $F_{xH^0} = N\sqrt{kT/2\pi m}$, where N and kT are the density and temperature of the Maxwell distribution, respectively, and m is the hydrogen mass. The upward flux of hydrogen ENAs calculated from the best fit parameters is $(1.13 \pm 0.18) \times 10^5\,\mathrm{cm^{-2}\,s^{-1}}$, whereas direct integration yields $(1.08 \pm 0.07) \times 10^5\,\mathrm{cm^{-2}\,s^{-1}}$. The best-fit flux error is from propagation of the fitting errors, while the integrated flux error is from propagation of the uncertainty owing to counting statistics.

I use the data from SWIM to calculate the downward proton flux impinging the lunar surface. I first conduct spacecraft potential correction. Spacecraft surfaces are charged negatively in the optical shadow owing to the larger flux of faster electrons than slower ions [26]. The ambient protons are attracted by the negative spacecraft potential before detected by SWIM, resulting in a shift in the energy spectrum toward higher energies. The Chandrayaan-1 spacecraft potential was not directly measured, but it is possible to constrain the upper and lower spacecraft potentials to be between 0 and −250 V, since the lower energy cutoff of the observed proton counts was around 250 eV during orbits 2563–2565. Figure 4.6 shows the proton energy spectra derived from the SWIM data with corrections of assumed spacecraft potentials of 0, −150 and −250 V. The energy spectra fitted by the Maxwell distribution and the best-fit parameters (the proton density and temperature) are also displayed.

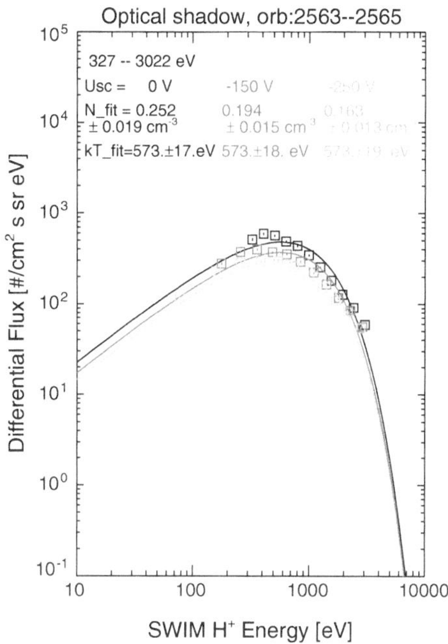

Fig. 4.6 Energy spectra of downward-traveling protons obtained by SWIM in the optical shadow of the Moon during orbits 2563–2565 for three cases of assumed spacecraft potentials of 0, −150 and −250 V. The energy spectra are shifted in units of distribution function and converted to differential flux. The shifted energy spectra are fitted by Maxwell distributions. The *solid curves* show the fitted distribution

Under the assumption of an isotropic proton distribution just above the Debye sheath formed on the charged lunar surface, the downward proton flux is calculated in the same way as calculation of the upward ENA fluxes. Since the Debye length above the lunar surface (<1 km) is much smaller than the lunar radius (1,738 km), acceleration by the negative lunar surface potential will not modify the downward proton flux onto the lunar surface from just above the sheath [26]. The downward proton fluxes calculated from the best fit parameters for the three assumed spacecraft potentials are $(2.35 \pm 0.23) \times 10^6\,\mathrm{cm^{-2}\,s^{-1}}$ for 0 V, $(1.81 \pm 0.18) \times 10^6\,\mathrm{cm^{-2}\,s^{-1}}$ for −150 V, and $(1.52 \pm 0.16) \times 10^6\,\mathrm{cm^{-2}\,s^{-1}}$ for −250 V, whereas direct integration yields $(2.13 \pm 0.02) \times 10^6\,\mathrm{cm^{-2}\,s^{-1}}$ for 0 V, $(1.70 \pm 0.01) \times 10^6\,\mathrm{cm^{-2}\,s^{-1}}$ for −150 V, and $(1.42 \pm 0.01) \times 10^6\,\mathrm{cm^{-2}\,s^{-1}}$ for −250 V. Using the fitted fluxes, I obtain backscattering ratios $r \sim 0.048 \pm 0.009$, 0.062 ± 0.012, 0.074 ± 0.014, and using the integrated fluxes $r \sim 0.051 \pm 0.003$, 0.064 ± 0.004, 0.076 ± 0.005 for the assumed spacecraft potentials of 0, −150, −250 V, respectively. Note that the two different methodologies result in only small difference in the hydrogen ENA/proton flux ratios. These backscattering ratios in the plasma sheet are comparable to, but slightly smaller than the solar wind backscattering ratio of ~0.1–0.2.

I now investigate the ENA backscattering ratios as functions of selenographic latitude using the downward proton fluxes and upward ENA fluxes derived from the SWIM and CENA measurements. During orbits 2563–2565, Chandrayaan-1 traveled at longitudes ~170°E, where a large magnetized region exists in the southern hemisphere (Fig. 4.4). Figure 4.7 shows backscattering ratios derived from the data

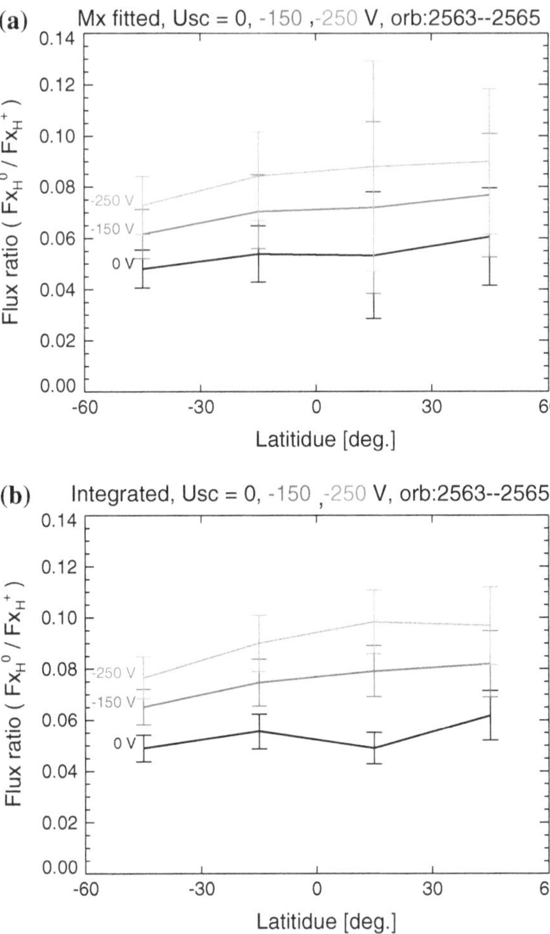

Fig. 4.7 Backscattering ratios along the selenographic longitude of ~180° as functions of selenographic latitude for **a** Maxwellian fitting and **b** direct integration to calculate the hydrogen ENA and proton fluxes. The data are taken from ±60° latitudes divided into four ranges. Three cases for spacecraft potentials of 0, −150, −250 V are shown. Error bars represent the error estimates from propagation of the fitting errors of the ENA and proton energy spectra

obtained in four different latitude ranges using the two flux calculation methods. Previous analysis of the CENA data obtained in the solar wind shows a substantial reduction in backscattering ratio up to ~50 % at the same longitude in the southern hemisphere at location of strong magnetization of the lunar surface [25]. The decrease in backscattering ratio can be attributed to the shielding effect of the lunar surface from the solar wind protons above the lunar magnetic anomalies. In the plasma sheet, on the other hand, the north-south asymmetry in the backscattering ratios is less clear for both of the two methods. Although one can see a tendency of smaller backscattering ratio in the southern hemisphere, the backscattering ratio difference between the north and south latitudes is at most ~20 %, which is much smaller than ~50 % decrease seen between −30° and −60° latitudes in the solar wind [25]. The smaller north-south asymmetry in the backscattering ratios suggests less effective shielding of the magnetized surface from the plasma sheet protons than the solar wind protons.

4.4 Magnetic Shielding from the Solar-Wind and Plasma-Sheet Protons

The observed difference in shielding efficiencies may be attributed to difference in velocity distributions of the solar wind and plasma sheet protons. In the Moon rest frame, the solar wind protons have a beam-like velocity distribution with a narrow thermal spread centered at the solar wind bulk velocity. On the other hand, the hot plasma sheet protons have a broad velocity distribution with a wide thermal spread. The proton thermal speed can become comparable to or larger than the bulk speed, depending on the geomagnetic conditions and Moon's location in the plasma sheet. The wide spread in velocity distributions may modify the proton dynamics around the lunar magnetic anomalies.

I conduct four sets of test-particle simulations including only magnetic force by a single dipole to investigate how the different velocity distributions of incident solar-wind and plasma-sheet protons ("mono-directional" and "isotropic" angular distributions combined with "mono-energy" and "Maxwellian" energy distributions) modify the magnetic shielding of the surface. I place a horizontal dipole along y-axis, buried at 20 km below the surface with a surface magnetic strength of 1000 nT. The maximum dipole-field strength at 100 km above the surface is 1.6 nT, which is larger than the strongest field strength of ~2.4 nT observed at 100 km altitude [22]. The lunar crustal magnetic fields actually have complex structures [13], and it is suggested that these higher-order terms can enhance the shielding efficiency without increasing the field strength at 100 km altitude [5]. The use of a rather strong magnetic field is to qualify the shielding effects by the crustal magnetic force under a simple setup of simulation (e.g., a single dipole). I set a simulation box with a 100 km altitude above the surface and a horizontal extent of $800 \times 800\,\mathrm{km}^2$ centered at the dipole location. I then launch a number of protons downward from the top of the box with uniformly distributed random starting locations, with a constant kinetic energy of 500 eV ("mono-energy" case) and a Maxwell energy distribution with a characteristic energy of 500 eV ("Maxwellian" case), and with two different initial velocity directions: a mono-directional distribution ("mono-directional" case) and an isotropic distribution ("isotropic" case). Note that the "mono-directional" case corresponds to the normal incidence of the solar wind protons to the subsolar lunar surface, whereas the "isotropic" case applies to any point on the lunar surface when the Moon is immersed in an isotropic plasma. I trace the particles until they strike the surface or escape from the simulation box. Figure 4.8a, b display sample proton trajectories for "mono-directional, mono-energy" and "isotropic, mono-energy" cases in the vicinity of the dipole ($200 \times 200\,\mathrm{km}^2$). If the particles strike the surface, the surface locations are recorded. After 4×10^6 particles are traced, I derive a shielding efficiency for each surface bin. Here I define the shielding efficiency as $SE = 1 - N_\mathrm{B}/N_\mathrm{noB}$, where N_B is a surface density of particles landing in a particular surface bin and N_noB is the average surface density of particles per surface bin obtained by another simulation without the dipole field. This shielding efficiency corresponds to the one used in the previous analysis [24]. Positive values of the shielding efficiency indicate that the incident flux to the surface is reduced at that point and negative values represent enhanced flux to the surface. I note that the backscattering ratio is related to the shielding efficiency as $r = r_\mathrm{noB}(1 - SE)$, where r_noB is the backscattering ratio at the surface where no crustal field exists. The surface distributions of the calculated shielding efficiencies for the different velocity distributions of launched protons are shown in Fig. 4.8c–f.

I first discuss the shielding efficiency distribution for the "mono-directional, mono-energy" proton case (corresponding to the "beam" solar wind protons). In Fig. 4.8c, one can see a large void region ($SE = 1$) formed mainly in the $-x$ region. This asymmetry in x-direction is caused by the negative B_y component above the surface, deflecting protons toward the $-x$-direction as shown in Fig. 4.8a. Figure 4.8c also shows an enhanced-flux region ($SE < 0$) around the shielded region associated with the protons that are deflected but still strike the surface. The pairs of void and enhanced-flux regions are common characteristics of the solar wind interaction with the lunar magnetic anomalies, which can be found in the previous observations and simulations in the solar wind [6, 7, 24, 27].

In contrast to the large void region caused by the coherent deflection of "mono-directional, mono-energy" protons shown in Fig. 4.8c, d shows different features of magnetic shielding for the "isotropic, mono-energy" proton case. A void region ($SE = 1$) is also formed just above the buried dipole, but its area is much smaller than that seen in the "mono-directional, mono-energy" proton case. No significant enhanced region is seen for the "isotropic, mono-energy" proton case. As shown in Fig. 4.8b, the "isotropic, mono-energy" protons enter the dipole field from random directions, resulting in incoherent deflection with no favored landing locations on the surface. Consequently, only the small area with strong horizontal fields is effectively shielded from the incident isotropic protons.

In addition to the angular spread, wider energy distributions of protons also modify the magnetic shielding of the surface. The enhanced flux region ($SE < 0$) seen for the "mono-directional, mono-energy" case (see Fig. 4.8c) is smeared out for the "mono-directional, Maxwellian" case except the magnetic cusp regions located at $(0, \pm 25)$ km on the surface (see Fig. 4.8e). For the Maxwellian case, the deflected lower- and higher-energy protons with both smaller and larger gyroradii strike wider regions on the surface compared to the coherently deflected mono-energy protons, forming the less clear boundary of the void region.

The difference of the "isotropic, Maxwellian" case (corresponding to the "broad" plasma sheet protons, Fig. 4.8f) from the "isotropic, mono-energy" case (Fig. 4.8d)

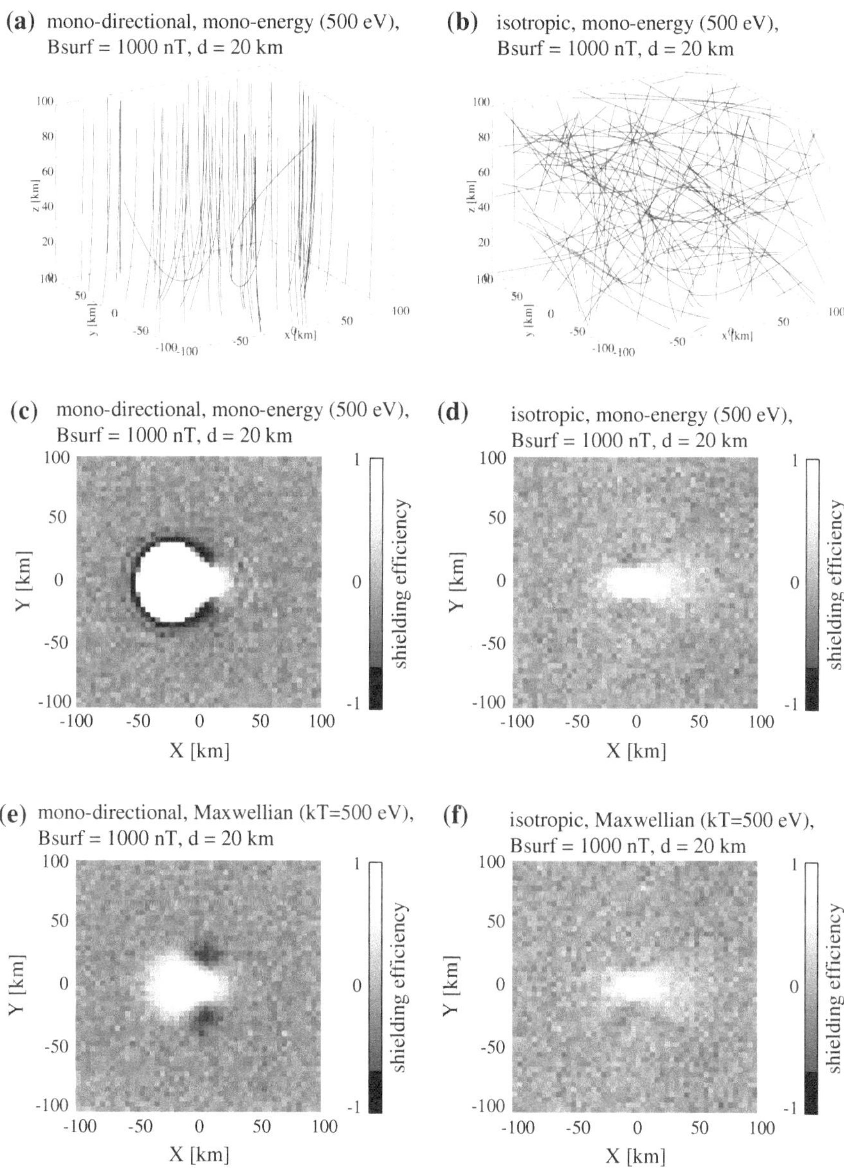

Fig. 4.8 Test particle trajectories around a buried dipole using different proton velocity distributions. A part of proton trajectories for **a** mono-directional and **b** isotropic initial velocity distributions with a constant energy of 500 eV, in addition to the shielding efficiencies on the surface (see text for details) for **c** mono-directional and **d** isotropic mono-energy protons (500 eV) and **e** mono-directional and **f** isotropic Maxwellian protons ($kT = 500$ eV) are shown. The dipole is horizontally placed along y-axis; the magnetic field has $-y$-component above the surface

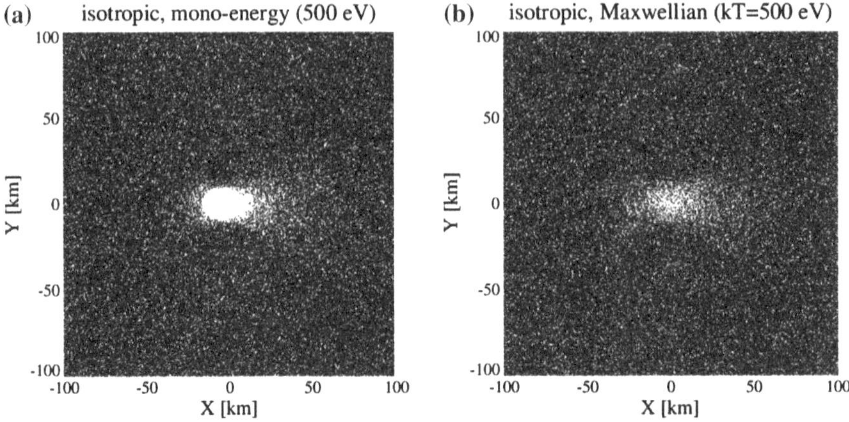

Fig. 4.9 Particle landing locations for the **a** "isotropic, mono-energy" and **b** "isotropic, Maxwellian" cases, which correspond to the shielding efficiency maps displayed in Fig. 4.8d, f, respectively

is small, but one can see that some of the white bins ($SE = 1$) in Fig. 4.8d turn into light gray ($SE < 1$) in Fig. 4.8f. Figure 4.9a, b display the particle landing locations instead of the shielding efficiencies for the "isotropic, mono-energy" and "isotropic, Maxwellian" cases. In this format, one can clearly see that some protons strike the "shielded" surface even just above the dipole for the Maxwellian case. This change can be attributed to high-energy protons capable of penetrating the dipole field. As I have presented, the combination of the angular and energy spread in proton VDFs prevents development of a large, clear void region above the buried magnetic dipole. The test particle simulations suggest that the surface is less effectively shielded by the dipole magnetic fields from the broad-VDF protons than from the beam-VDF protons, as inferred from the CENA observations in the plasma sheet and the solar wind.

I note that this model neglects any collective effects and focuses on the deflection of single particles by the magnetic force, as an early model of solar-wind shielding does [6]. The lack of collective effects hinders quantitative evaluation of the shielding efficiency. For example, charge-separation electric fields will enhance the ion deflection, wheares plasma compression of crustal magnetic fields may result in less effective shielding because of the decrease in scale size length of the mini-magnetosphere. In addition, a bow shock or shock-like formation associated with a mini-magnetosphere in the solar wind might play a role in protecting the lunar surface from the impinging protons. The difference in potential distributions and magnetic-field configurations above the magnetic anomalies resulting from the smaller pressure gradient and lower Mach numbers in the plasma sheet than in the solar wind will affect the upward ENA flux from the Moon and will be examined in future study. Nevertheless, this model can qualitatively exhibit the basic characteristics of magnetic shielding of the surface from the "beam" and "broad" ions.

4.5 Conclusions

In this chapter, I present observations of ENAs from the lunar surface in the terrestrial plasma sheet. The observed ENA energy spectra are well-fitted by the Maxwell distribution with densities of $\sim$0.03 cm^{-3} and characteristic energies (temperatures) of $\sim$100 eV. The hydrogen ENA/proton flux ratios $\sim$0.05–0.08 are roughly on the same order of magnitude of the backscattering ratios of the solar wind protons $\sim$0.1–0.2. These characteristics are similar to the backscattered hydrogen ENAs in the solar wind, suggesting that the detected ENAs are plasma sheet particles backscattered as hydrogen ENAs from the lunar surface.

On the other hand, the flux ratio of the backscattered hydrogen ENAs to the downward plasma sheet protons observed in the different selenographic latitudes exhibits no significant difference in the northern and southern hemispheres. No substantial reduction of the backscattering ratio of plasma sheet protons is found over a large magnetic anomaly region at selenographic longitudes of $\sim$170°E in the southern hemisphere, while the solar wind backscattering ratio drops to about the half in the southern hemisphere compared to that in the northern hemisphere. This result suggests a smaller shielding efficiency by lunar magnetic anomalies in the plasma sheet than in the solar wind. Test-particle simulations consistently show less effective shielding in the case of "broad" incident protons than that in the "beam"-proton case. The simulations suggest that difference between the beam-like velocity distributions of the solar wind protons and the broad velocity distributions of the plasma sheet protons may introduce different behavior of incident protons around the magnetic anomalies. Since this model only considers the dipole magnetic force on single particle motion, I have discussed the ion shielding effect in a mostly qualitative manner. More sophisticated models that can take into account collective effects such as magnetic-field deformation by plasma currents and charge-separation electric fields are necessary to evaluate the shielding efficiencies quantitatively and to understand the physical processes actually working around the magnetic anomalies. It would be also important to include more realistic configurations (e.g., higher-order terms) of lunar crustal magnetic fields [3, 5]. As I have shown for lunar ENAs, backscattered ENA observations within planetary magnetospheres can extend our knowledge on the solar-wind interactions with solid surfaces and crustal magnetic fields to a completely different plasma regime.

References

1. Barabash, S., Bhardwaj, A., Wieser, M., Sridharan, R., Kurian, T., Varier, S., Vijayakumar, E., Abhirami, V., Raghavendra, K.V., Mohankumar, S.V., Dhanya, M.B., Thampi, S., Kazushi, A., Andersson, H., Yoshifumi, F., Holmström, M., Lundin, R., Svensson, J., Karlsson, S., Piazza, R.D., Wurz, P.: Investigation of the solar wind-Moon interaction onboard Chandrayaan-1 mission with the SARA experiment. Curr. Sci. **96**(4), 526–532 (2009)

2. Futaana, Y., Barabash, S., Wieser, M., Holmström, M., Lue, C., Wurz, P., Schaufelberger, A., Bhardwaj, A., Dhanya, M.B., Asamura, K.: Empirical energy spectra of neutralized solar wind protons from the lunar regolith. J. Geophys. Res. **117**, E05005 (2012). doi:10.1029/2011JE004019
3. Harada, Y., Machida, S., Saito, Y., Yokota, S., Asamura, K., Nishino, M.N., Tsunakawa, H., Shibuya, H., Takahashi, F., Matsushima, M., Shimizu, H.: Small-scale magnetic fields on the lunar surface inferred from plasma sheet electrons. Geophys. Res. Lett. **40**(13), 3362–3366 (2013). doi:10.1002/grl.50662. http://dx.doi.org/10.1002/grl.50662
4. Harnett, E.M., Cash, M., Winglee, R.M.: Substorm and storm time ionospheric particle flux at the moon while in the terrestrial magnetosphere. Icarus **224**(1), 218–227 (2013). doi:10.1016/j.icarus.2013.02.022. http://www.sciencedirect.com/science/article/pii/S0019103513O
5. Harnett, E.M., Winglee, R.M.: 2.5-D Fluid simulations of the solar wind interacting with multiple dipoles on the surface of the Moon. J. Geophys. Res. **108**, 1088 (2003). doi:10.1029/2002JA009617
6. Hood, L.L., Williams, C.R.: The lunar swirls—distribution and possible origins.In: Proceedings of Lunar planetery sciience conference, 19th, pp. 99–113 (1989)
7. Kallio, E., Jarvinen, R., Dyadechkin, S., Wurz, P., Barabash, S., Alvarez, F., Fernandes, V.A., Futaana, Y., Harri, A.M., Heilimo, J., Lue, C., Mäkelä, J., Porjo, N., Schmidt, W., Siili, T.: Kinetic simulations of finite gyroradius effects in the lunar plasma environment on global, meso, and microscales. Planet. Space Sci. **74**, 146–155 (2012). doi:10.1016/j.pss.2012.09.012
8. Kazama, Y., Barabash, S., Wieser, M., Asamura, K., Wurz, P.: Development of an LENA instrument for planetary missions by numerical simulations. Planet. Space Sci. **55**(11), 1518–1529 (2007). doi:10.1016/j.pss.2006.11.027
9. Machida, S., Mukai, T., Saito, Y., Hirahara, M., Obara, T., Nishida, A., Terasawa, T., Maezawa, K.: Plasma distribution functions in the Earth's magnetotail ($X_{GSM} \sim -42R_E$) at the time of a magnetospheric substorm: GEOTAIL/LEP observation. Geophys. Res. Lett. **21**, 1027–1030 (1994). doi:10.1029/94GL00187
10. Matsushima, M., Tsunakawa, H., Iijima, Y.i., Nakazawa, S., Matsuoka, A., Ikegami, S., Ishikawa, T., Shibuya, H., Shimizu, H., Takahashi, F.: Magnetic cleanliness program under control of electromagnetic compatibility for the SELENE (Kaguya) spacecraft. Space Sci. Rev. **154**(1–4), 253–264 (2010). doi:10.1007/s11214-010-9655-x
11. McCann, D., Barabash, S., Nilsson, H., Bhardwaj, A.: Miniature ion mass analyzer. Planet. Space Sci. **55**(9, Sp. Iss. SI), 1190–1196 (2007). doi:10.1016/j.pss.2006.11.020
12. Mukai, T., Yamamoto, T., Machida, S.: Dynamics and kinetic properties of plasmoids and flux ropes: GEOTAIL observations. Geophys. Monogr. Ser. **105**, 117–137 (1998). doi:10.1029/GM105p0117
13. Purucker, M.E.: A global model of the internal magnetic field of the Moon based on Lunar Prospector magnetometer observations. Icarus **197**(1), 19–23 (2008). doi:10.1016/j.icarus.2008.03.016
14. Rich, F.J., Reasoner, D.L., Burke, W.J.: Plasma sheet at Lunar distance: characteristics and interactions with the Lunar surface. J. Geophys. Res. **78**(34), 8097–8112 (1973). doi:10.1029/JA078i034p08097
15. Rodríguez M., D., Saul, L., Wurz, P., Fuselier, S., Funsten, H., McComas, D., Möbius, E.: IBEX-Lo observations of energetic neutral hydrogen atoms originating from the lunar surface. Planet. Space Sci. **60**(1), 297–303 (2012). doi:10.1016/j.pss.2011.09.009. http://www.sciencedirect.com/science/article/pii/S0032063311O
16. Saito, Y., Nishino, M., Yokota, S., Tsunakawa, H., Matsushima, M., Takahashi, F., Shibuya, H., Shimizu, H.: Night side lunar surface potential in the Earth's magnetosphere. Adv. Space Res. (2013). doi:10.1016/j.asr.2013.05.011. http://www.sciencedirect.com/science/article/pii/S027311771 3O
17. Saito, Y., Yokota, S., Asamura, K., Tanaka, T., Akiba, R., Fujimoto, M., Hasegawa, H., Hayakawa, H., Hirahara, M., Hoshino, M., Machida, S., Mukai, T., Nagai, T., Nagatsuma, T., Nakamura, M., ichiro Oyama, K., Sagawa, E., Sasaki, S., Seki, K., Terasawa, T.: Low-energy charged particle measurement by MAP-PACE onboard SELENE. Earth Planets Space **60**, 375–385 (2008)

18. Saito, Y., Yokota, S., Asamura, K., Tanaka, T., Nishino, M.N., Yamamoto, T., Terakawa, Y., Fujimoto, M., Hasegawa, H., Hayakawa, H., Hirahara, M., Hoshino, M., Machida, S., Mukai, T., Nagai, T., Nagatsuma, T., Nakagawa, T., Nakamura, M., ichiro Oyama, K., Sagawa, E., Sasaki, S., Seki, K., Shinohara, I., Terasawa, T., Tsunakawa, H., Shibuya, H., Matsushima, M., Shimizu, H., Takahashi, F.: In-flight performance and initial results of plasma energy angle and composition experiment (PACE) on SELENE (Kaguya). Space Sci. Rev. **154**, 265–303 (2010). doi:10.1007/s11214-010-9647-x
19. Shimizu, H., Takahashi, F., Horii, N., Matsuoka, A., Matsushima, M., Shibuya, H., Tsunakawa, H.: Ground calibration of the high-sensitivity SELENE Lunar magnetometer LMAG. Earth Planets Space **60**, 353–363 (2008)
20. Takahashi, F., Shimizu, H., Matsushima, M., Shibuya, H., Matsuoka, A., Nakazawa, S., Iijima, Y., Otake, H., Tsunakawa, H.: In-orbit calibration of the lunar magnetometer onboard SELENE (KAGUYA). Earth Planets Space **61**, 1269–1274 (2009)
21. Troshichev, O., Kokubun, S., Kamide, Y., Nishida, A., Mukai, T., Yamamoto, T.: Convection in the distant magnetotail under extremely quiet and weakly disturbed conditions. J. Geophys. Res. **104**(A5), 10249–10263 (1999). doi:10.1029/1998JA900141. http://dx.doi.org/10.1029/1998JA900141
22. Tsunakawa, H., Shibuya, H., Takahashi, F., Shimizu, H., Matsushima, M., Matsuoka, A., Nakazawa, S., Otake, H., Iijima, Y.: Lunar magnetic field observation and initial global mapping of Lunar magnetic anomalies by MAP-LMAG onboard SELENE (Kaguya). Space Sci. Rev. **154**, 219–251 (2010). doi:10.1007/s11214-010-9652-0
23. Tsunakawa, H., Takahashi, F., Shimizu, H., Shibuya, H., Matsushima, M.: Regional mapping of the lunar magnetic anomalies at the surface: method and its application to strong and weak magnetic anomaly regions. Icarus **228**(0), 35–53 (2014). doi:10.1016/j.icarus.2013.09.026. http://www.sciencedirect.com/science/article/pii/S00191035130
24. Vorburger, A., Wurz, P., Barabash, S., Wieser, M., Futaana, Y., Holmström, M., Bhardwaj, A., Asamura, K.: Energetic neutral atom observations of magnetic anomalies on the lunar surface. J. Geophys. Res. **117**, A07208 (2012). doi:10.1029/2012JA017553
25. Vorburger, A., Wurz, P., Barabash, S., Wieser, M., Futaana, Y., Lue, C., Holmström, M., Bhardwaj, A., Dhanya, M.B., Asamura, K.: Energetic neutral atom imaging of the Lunar surface. J. Geophys. Res. **118**, 3937–3945 (2013). doi:10.1002/jgra.50337. http://dx.doi.org/10.1002/jgra.50337
26. Whipple, E.C.: Potentials of Surfaces in Space. Rep. Prog. Phys. **44**(11), 1197–1250 (1981). doi:10.1088/0034-4885/44/11/002. http://stacks.iop.org/0034-4885/44/i=11/a=002
27. Wieser, M., Barabash, S., Futaana, Y., Holmström, M., Bhardwaj, A., Sridharan, R., Dhanya, M.B., Schaufelberger, A., Wurz, P., Asamura, K.: First observation of a mini-magnetosphere above a Lunar magnetic anomaly using energetic neutral atoms. Geophys. Res. Lett. **37**, L05103 (2010). doi:10.1029/2009GL041721

Chapter 5
Conclusions

I have analyzed the plasma, neutral-atom, and electromagnetic-field data from three lunar missions: Kaguya, ARTEMIS, and Chandrayaan-1. I particularly focus on the near-lunar environment in the Earth's magnetotail. Interactions of the Moon's surface, plasma, and magnetic anomalies with the terrestrial magnetotail plasma exhibit remarkable Moon-related features in a completely different plasma regime from the solar wind. The comparison of Moon–plasma interactions in the Earth's magnetotail with Moon-solar wind interactions can give insight into general processes at work.

First I have shown Kaguya observations of kinetic electron behavior with spatial scale lengths comparable to the electron gyroradius. Kaguya observed non-gyrotropic electron velocity distribution functions (VDFs) associated with the gyro-loss effect, namely electron absorption by the lunar surface combined with electron gyromotion. I conduct particle-trace calculations to derive theoretical forbidden regions in the electron VDFs, thereby taking into account the modifications by diamagnetic-current systems, lunar-surface charging, and perpendicular electric fields. Comparison between the observed empty regions with the theoretically derived forbidden regions suggests that several components can modify the characteristics of the non-gyrotropic electron VDFs depending on the ambient-plasma conditions. On the lunar nightside in the Earth's magnetotail lobes, negative surface potentials slightly reduce the size of the forbidden regions, but there are no distinct effects of either the diamagnetic current or perpendicular electric fields. On the dayside in the solar wind, the observations suggest the presence of either the diamagnetic-current or solar-wind motional electric-field effects, or both. In the terrestrial plasma sheet, all three mechanisms can substantially modify the characteristics of the forbidden regions. The observations in the plasma sheet imply the presence of a local electric field of at least 5 mV/m although the mechanism responsible for production of such a strong electric field is unknown. In addition, I have presented an application of the kinetic behavior of electrons near the Moon to remote measurements of small-scale magnetic fields on the lunar surface. The origins of the lunar crustal magnetic fields remain unclear although dozens of magnetic field measurements have been conducted on and above the lunar surface. A major obstacle to resolving this problem is the extreme difficulty of determining a surface distribution of small-scale magnetization.

Y. Harada, *Interactions of Earth's Magnetotail Plasma with the Surface, Plasma, and Magnetic Anomalies of the Moon*, Springer Theses,
DOI 10.1007/978-4-431-55084-6_5

I developed a new technique to map small-scale magnetic fields using non-adiabatic scattering of high-energy electrons in the terrestrial plasma sheet. Particle tracing, utilizing three-dimensional lunar magnetic field data synthesized from magnetometer measurements, enables separation of the contributions to electron motion of small- and large-scale magnetic fields. The obtained map shows significant kilometer-scale magnetic fields on the southwestern side of the South Pole–Aitken basin that are correlated with larger-scale magnetization. This implies that kilometer-scale magnetization may be present over the lunar surface and related to the large-scale magnetization. As is the case for the lunar surface, analysis of non-gyrotropic VDFs near solid surfaces in space can promote a better understanding of the near-surface electromagnetic environment and of plasma–solid-surface interactions.

I have demonstrated an important role of plasma of lunar origin in the Earth's magnetotail lobes from observations by the dual-probe ARTEMIS mission of Moon-related electron and ion signatures obtained above the dayside lunar surface. While the Moon is often thought of as a passive absorber, recent observations indicate that plasma of lunar origin can have significant effects on the near-lunar environment. I present new observations from ARTEMIS showing that lunar plasma can play a substantial role in the low-density environment of the terrestrial magnetotail. Two-point observations reveal that the density of plasma of lunar origin can be higher than that of the ambient lobe plasma even several hundreds of kilometers above the Moon's dayside. Meanwhile, the distributions of incoming lobe electrons exhibit modifications correlated with the presence of the Moon-related populations, suggesting direct or indirect interactions of the lobe electrons with plasma of lunar origin. I also show observations of high-energy photoelectron emission from the dayside lunar surface, supporting the existence of large positive potentials on the lunar surface. Lunar pickup ions with nonzero parallel-velocity components provide further evidence for positive surface potentials of tens of volts or more. ARTEMIS data reveal not only the existence of the positive surface potentials much larger than those predicted from a current-balance model based on Maxwellian plasmas, but also their significant implications for the dynamics of both the dominant Moon-originating ions and the tenuous ambient plasma populations in the terrestrial magnetotail lobes. The high-energy photoelectrons and interaction between plasmas of ambient and exospheric origins may be important at other airless bodies as well.

Finally I have presented the observations of energetic neutral atoms (ENAs) produced at the lunar surface in the Earth's plasma sheet. When the Moon was located in the plasma sheet, Chandrayaan-1 detected hydrogen ENAs upcoming from the Moon. Analysis of the hydrogen ENA and proton data reveal the characteristic energy of the observed ENA energy spectrum (the e-folding energy of the distribution function) $\sim$100 eV and the ratio of upward ENA flux to downward proton flux $<\sim 0.1$. These characteristics are similar to those of the backscattered ENAs observed in the solar wind, suggesting that Chandrayaan-1 detected plasma sheet particles backscattered as ENAs from the lunar surface. The hydrogen ENA/proton flux ratio observed in the plasma sheet exhibits no significant difference in the southern hemisphere, where a large and strong magnetized region exists, compared with the northern hemisphere. This is contrary to the previous observations in the solar wind, where the flux ratio in the southern hemisphere drops to $\sim$50 % of that in the northern hemisphere. These ENA observations and test-particle simulations suggest that magnetic shielding of the lunar surface in the plasma sheet is less effective than in the solar wind owing to the broad velocity distributions of the plasma sheet protons. Comparative studies of mini-magnetospheres in a wide range of plasma parameters can contribute to magnetospheric physics in general.

The new findings presented in this thesis regarding the near-lunar environment in the Earth's magnetotail have expanded our knowledge on the fundamental aspects of the Moon-plasma interactions. The near-lunar observations of plasma, electric and magnetic fields, and ENAs in the Earth's magnetotail with a wide range of ambient plasma parameters not only complement our current understanding of the Moon-solar wind interactions, which have been extensively studied for decades, but also provide implications for other space and planetary environments. As I have shown, the electron gyro-scale dynamics becomes important mainly at low altitudes ($<\sim$100 km) around the Moon. On the other hand, the electron kinetic effects may control the global nature of a plasma wake formed behind small airless bodies such as asteroids and distant comets if the body size is comparable to or smaller than the electron gyroradius. Non-gyrotropic electron VDFs will be present around these bodies as well. Kinetic treatment of electron dynamics will be necessary to study plasma processes that are active around small-scale airless bodies.

Non-adiabatic magnetic scattering of high-energy charged particles will also be important at other small-scale magnetic fields such as Mercury's magnetosphere interacting with the solar wind, Ganymede's magnetosphere immersed in jovian magnetospheric plasma, and Martian crustal magnetic fields exposed to both the solar wind and planetary plasma. Particle tracing is a basic and useful tool to study the charged-particle motion in these small-scale magnetic fields, though self-consistent treatment of particles and fields is required to include collective effects. Observations of backscattered ENAs from solid surfaces will provide valuable information about ion dynamics in the magnetospheres of Mercury and Ganymede.

Although atmospheric and ionospheric effects are often ignored when studying plasma interactions with "airless" bodies, one must exercise caution because plasma of surface or exospheric origin can play a dominant role in a tenuous plasma environment. Similar phenomena to the lobe electron modification associated with the presence of plasma of lunar origin might operate at other places. We do not yet fully understand how the newly ionized plasma interacts with the ambient plasma in a wide variety of plasma regimes. Plasma of surface and exospheric origin should be incorporated in a framework of airless-body plasma physics.